Osnei Francisco Alves

Socio-economic Analysis of a Biodiesel Plant

Osnei Francisco Alves

Socio-economic Analysis of a Biodiesel Plant

Biodiesel Technology and Production in the State of Maranhão - Brazil

ScienciaScripts

Cover image: www.ingimage.com

This book is a translation from the original published under ISBN 978-613-9-61385-4.

Publisher:
Sciencia Scripts
is a trademark of
Dodo Books Indian Ocean Ltd. and OmniScriptum S.R.L publishing group

120 High Road, East Finchley, London, N2 9ED, United Kingdom
Str. Armeneasca 28/1, office 1, Chisinau MD-2012, Republic of Moldova, Europe
Printed at: see last page
ISBN: 978-620-7-60900-0

Table of contents:

ACKNOWLEDGMENTS

To God.

To Professor and advisor Dr. Helena Maria Wilhelm, for her competence, availability, patience, criticism and relevant suggestions during her guidance.

To the coordinator, Dr. Maurício Cantão, and all the Prodetec professors for their knowledge and competence.

The Graduate Program of the Institute of Technology for Development (LACTEC) and the Paraná Engineering Institute (IEP).

To my colleagues in the Master's program.

My wife Ursula da Silva Alves and my son Vinícius Francisco Alves.

To my colleagues at Faculdades Santa Cruz and Banco CNH Capital.

SUMMARY

The aim of this work was to carry out a socio-economic analysis of the implementation of a biodiesel plant in the state of Maranhão, taking into account technical, economic and socio-environmental concepts. In order to set up the biodiesel plant in Maranhão, a survey was carried out of the main raw materials available, with a view to selecting the most promising raw materials, as well as favorable regions for biodiesel production. Afterwards, the available workforce and the existence of cooperatives in the potential biodiesel plant installation region were analyzed. Next, an economic analysis of biodiesel production was carried out and, at the end, the socio-economic impact on the cooperative members of setting up a biodiesel plant in the region was estimated. In addition, based on fixed and variable costs, financial income and payback, within the forecast of sales of biodiesel and its by-products, the enterprise was analyzed and simulated to fit into the PRONAF credit line. Based on the results obtained, it was possible to conclude that the biodiesel production plant using babassu oil in the Médio Mearim region is socio-economically viable, considering the sale of biodiesel and by-products from the babassu coconut.

Key words: Babassu. Biodiesel. Cooperatives. Sustainable development.

Chapter 1

1. INTRODUCTION

Despite the recent prominence that the energy use of vegetable oils has received in the national and international media, this is an old idea, which was first suggested at the end of the 19th century when Rudolph Diesel, inventor of the internal compression combustion engine (Diesel cycle engine), used petroleum, alcohol and peanut oil as fuels in his tests (KNOTHE et al , 2005). However, due to the low cost and high availability of oil at the time, this fossil raw material became the world's main source of liquid fuels.

Currently, most of the energy consumed in the world comes from oil-derived sources, with diesel oil being one of its main fractions. However, at various times in the 20th century, especially in supply crises during the two world wars, fresh vegetable oils were strategically used as liquid fuels (MA and HANNA, 1999).

Finally, the crisis in the world oil market in the 1970s led to a movement towards the production of alternative liquid fuels from renewable sources. At this time, a technological alternative was proposed that allowed the viscosity of vegetable oils to be reduced, bringing their characteristics closer to those of diesel oil and facilitating their use as a fuel for diesel cycle engines. This reaction is known as the transesterification or alcoholysis of vegetable oils and gives rise, as a reaction product, to a mixture of alkyl monoesters, such as methyl or ethyl esters, known as "BIODIESEL".

This mixture was initially called "PRODIESEL" in Brazil. The studies that led to the development of PRODIESEL, from different sources of vegetable oils such as soybean, babassu, peanut, cotton and sunflower, began in the 1970s at the Federal University of Ceará (PARENTE, 2003).

It is now widely recognized that the availability of oil and its derivatives on the world market is finite, and excessive dependence on its supply poses serious socio-economic and environmental problems. Therefore, the creation and maintenance of programs aimed at researching alternative sources of renewable energy, with the aim of totally or partially replacing fossil fuels, have been given great priority in recent decades (KNOTHE et al, 2005; RAMOS and WILHELM, 2005), and these initiatives are of vital importance for developing economies.

Among the most suitable and available biomass sources for consolidating renewable energy programs, vegetable oils have been investigated not only for their properties, but also because they represent alternatives for decentralized energy generation, acting as strong support for family farming, creating better living conditions in poor regions, enhancing regional potential and offering alternatives to economic and socio-environmental problems.

1.1 BACKGROUND

Worldwide pressure on the use of non-renewable fuels has led some industrialized countries to set targets for reducing gas emissions into the atmosphere (Kyoto Convention), committing themselves to looking for new sources of alternative energy. Examples of these alternative energies to the use of petroleum diesel include biodiesel, a biofuel produced from the reaction of vegetable oils or animal fats with alcohols such as methanol and ethanol (PARENTE, 2003).

It is estimated that 80% of all energy consumed by humanity comes from fossil fuel derivatives. The incessant use of these fuels releases particulate carbon or soot and toxic gases such as carbon monoxide and dioxide, sulphur oxide and polycyclic aromatic hydrocarbons into the atmosphere (LIMA, 2005).

A recurring concern in discussions about the scale and location of industrial biodiesel production units is that the product can be produced on a small scale, encouraging the use of raw materials located in less favored regions of the country. Within this concept is the prioritization of oilseeds that contribute to increasing employment and that are preferably grown in regions that are on the margins of the economic development process.

The state of Maranhão is seeking to develop actions and projects focused on environmental and technological issues that can guarantee the sustainable development of the region, as shown in Figure 1. Along these lines, the state has a great deal of plant biodiversity for the development of biofuels that can potentially contribute to socio-environmental development, raising the income of the local population.

Among the various native plant species found in the northeastern region of the country, more specifically in Maranhão, the focus of this work, are the palm families such as babassu, dendê, carnauba, etc.

The main objective of this research was to assess the socio-economic viability of setting up a biodiesel production plant in Maranhão, based on local extractivism and small farmers with no profitable alternatives, organized in the form of a cooperative.

In order to achieve the main objective, the research was divided into two phases. In the first, a socio-economic analysis of the state of Maranhão was carried out, and in the second, an economic analysis of biodiesel production based on the potential raw material and the location of the plant defined in the first phase.

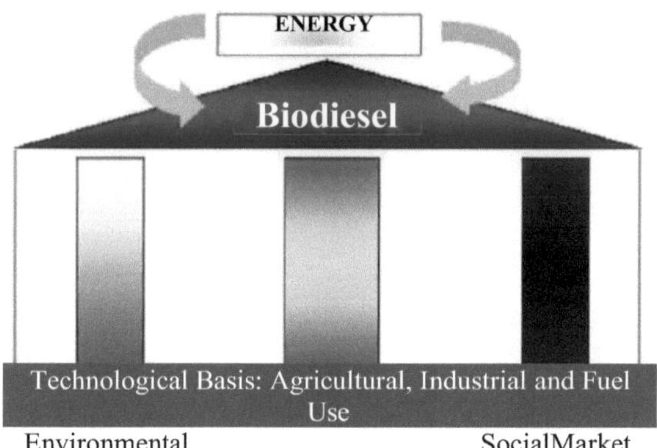

Figure 1: Pillars of the Biodiesel Production and Use Program in Brazil - PNPB

Source: BIODIESEL, 2009

1.2 OBJECTIVES

1.2.1 General Objective

A socio-economic analysis of the feasibility of setting up a biodiesel production plant in the state of Maranhão, based on the selection of a potential regional raw material and small farmers organized in the form of a cooperative.

1.2.2 Specific objectives

Select and evaluate the availability and location of potential raw materials for biodiesel production in Maranhão.

Calculate the estimated production of biodiesel from the selected raw material.

Dimension the size of the biodiesel plant and estimate the cost of installing it.

Analyze the available workforce and the existence of cooperatives in the region selected for the biodiesel production plant.

Estimate the socio-economic impact on the cooperative's members of setting up the biodiesel plant in the region previously selected.

Chapter 2

2. LITERATURE REVIEW

2.1 DEFINITION AND CHARACTERISTICS OF BIODIESEL

Biodiesel has been defined by the National Biodiesel Board of the United States as a mono-alkyl ester derivative of long-chain fatty acids from renewable sources such as vegetable oils or animal fat, whose use is associated with the replacement of fossil fuels in compression ignition engines (NATIONAL BIODIESEL BOARD, 1999).

According to Law No. 11.097, of January 13, 2005, biodiesel can be classified as any alternative fuel, of a renewable nature, which can offer socio-environmental advantages when used as a total or partial substitute for petroleum diesel in internal compression ignition engines (diesel cycle engines). However, the only type of alternative biofuel already regulated in Brazil is alkyl esters derived from vegetable oils or animal fats, obtained from their reaction with an alcohol in the presence of a catalyst (transesterification reaction). In other words, biodiesel.

As a fuel, alkyl esters, or biodiesel, must meet the parameters set out in technical specifications, expressed in technical standards such as ASTM D6751 (American Standard Testing Methods, 2003), DIN 14214 (Deutsches Institut für Normung, 2003) or ANP Resolution n° 07 (Agência Nacional do Petróleo, 2008), which establishes the properties that the product must have (quality) in order to be accepted and sold on the Brazilian market, without causing damage to the engine (WILHELM et al, 2008) (Table 1).

The great compatibility of biodiesel with conventional diesel makes it an alternative capable of serving most of the diesel vehicle fleet already on the market, without any need for technological investment in engine development (WILHELM et al, 2008).

The advantages of vegetable oil as a fuel over diesel are: natural liquid, renewable, high energy value, low sulphur content, low aromatic content and biodegradable (FANGRUI et al., 1999).

Biodiesel also has important characteristics, such as greater viscosity and a higher flash point than conventional diesel. It is practically free of sulphur and aromatic compounds, has an average oxygen content of around 11% and produces emissions with lower levels of carbon dioxide (CO2), particulate matter and carbon monoxide (CO) (MA and HANNA, 1999; VAN GERPEN and KNOTHE, 2005).

Table 1. Preliminary specifications for biodiesel in Brazil

Properties	Limits	Methods
Flash point (°C)	100 min.	NBR14598; ISO/CD3679
Kinematic viscosity at 40°C (mm^2 /s)	3,0 - 6,0	NBR10441; D445; EN/ISO3104
Water content (mg/kg)	500.0 max.	ASTM D 6304; EN ISO 12937
Sulphated ash (%, m/m)	0.02 max.	NBR 9842; D874; IS3987
Total sulfur (mg/kg)	50.0 max.	D5453; EN/ISO14596
Corrosivity to copper for 3 hours at 50°C	No. 1 max.	NBR14359; D130; EN/ISO2160
Ester content (%, m/m)	96.5 min.	NBR 15342; EN 14103
Cetane number	Write it down	D613; EN/ISO5165
Carbon residue (%, m/m)	0.05 max.	D4530; EN/ISO10370
Acidity index (mg KOH/g)	0.50 max.	NBR14448; D664; prEN14104
Free glycerin (%, m/m)	0.02 max.	D6854; prEN14105-6
Total glycerin (%, m/m)	0.25 max.	D6854; prEN14105
Phosphorus (mg/kg)	Write it down	D4951; prEN14107
Specific mass at 20°C (kg/m)^3	850 - 900	NBR7148/14065; D1298/4052
Methanol or Ethanol, max. (%, m/m)	0.20 max.	NBR 15343; prEN14110
Iodine number	Write it down	prEN14111
Monoglycerides (%, m/m)	Write it down	D6584; prEN14105
Diglycerides (%, m/m)	Write it down	D6584; prEN14105
Triglycerides (%, m/m)	Write it down	D6584; prEN14105
Sodium and potassium (mg/kg)	5 max.	prEN14108-9
Oxidation stability at 110 °C (h)	6 min.	prEN14112

Source: ANP, 2008

The renewable nature of biodiesel is based on the fact that it is derived from raw materials from agricultural activities, unlike petroleum derivatives. The choice of ethanol as the transesterification agent (thus obtaining ethyl esters) instead of methanol makes biodiesel a totally renewable product, but interest in this idea is limited to regions where the commercial value of ethanol is compatible with the market value of petrochemical methanol. This situation is only found in countries that produce ethanol in volumes compatible with demand, as is the case in Brazil, and for this reason, studies into ethyl transesterification are not as advanced as those associated with the methyl route (RAMOS et al., 2003; RAMOS and WILHELM, 2005).

Among the various oilseeds known in the literature, those with a high oil content in the seed are favorable for biodiesel production. These include the oilseeds soybean, peanut, sunflower, babassu, corn, rapeseed, castor bean and cotton (VARGAS et al., 1999).

The introduction of ethyl biodiesel on the domestic market, as it involves the participation of various segments of society, such as the ethanol and oilseed production chains, opens up opportunities for major social benefits resulting from the high rate of job creation, culminating in the valorization of the countryside and the promotion of rural workers. It also fosters the demand for skilled labor in the processing of vegetable oils, allowing integration, where necessary, between small producers and large companies.

From an economic point of view, the viability of biodiesel is related to the establishment of a favorable balance in the Brazilian trade balance, given that diesel is the most consumed petroleum derivative in Brazil, and that an increasing fraction of this product is being imported every year.

In environmental terms, the adoption of biodiesel, even if progressively, i.e. starting with additions of 2% (B2) to 5% (B5) to petroleum diesel, will result in a significant reduction in the pattern of emissions of particulate materials, sulphur oxides and gases that contribute to the greenhouse effect (MITTELBACH et al, 1985).

In the long term, this will lead to longer life expectancies for the population and, as a consequence, a decline in public health spending, making it possible to direct funds to other sectors, such as education and social security. The addition of biodiesel can also improve the properties of petrodiesel, allowing for a reduction in engine noise levels, improving lubricity (particularly in low-sulphur fuels) and increasing combustion efficiency by increasing the cetane number (WILHELM et al , 2008).

It should also be noted that the inclusion of biodiesel in the national energy matrix represents a powerful element of synergy with the sugarcane agribusiness, the effect of which is extremely beneficial for the national economy (RAMOS, 1999 and 2003). Ethanol production is significant in practically all regions of the country, and has contributed to increasing the sector's competitiveness, making use of the existing distribution network and the excellent performance of the technologies developed for the sugarcane production chain. In this context, Brazil is in a position that no country has ever been in the history of the globalized world. With the obvious decline of fossil fuels, no other tropical region has the size and conditions to be one of the main suppliers of biofuels and clean technologies for the 21st century (WILHELM et al, 2008).

Although favorable from an energy point of view, the direct use of vegetable oils in diesel engines is very problematic. Studies carried out on various vegetable oils have shown that their direct combustion leads to a series of problems: carbonization in the injection chamber, resistance to ejection in the piston segments, dilution of the crankcase oil, contamination of the lubricating oil, among other problems. The causes of these problems have been attributed to the polymerization of triglycerides through their double bonds, which leads to the formation of deposits. Low volatility and high viscosity are also the main reasons why vegetable oils or fats are transesterified into biodiesel, as high viscosity leads to problems with fuel atomization (KNOTHE and STEIDLEY, 2005).

2.1.1 Biodiesel production process using the transesterification reaction

Biodiesel, fatty acid alkyl esters, has recently become very attractive due to its environmental benefits and the fact that it is produced from renewable resources. Currently, biodiesel is made commercially by transesterifying oils or fats with an alcohol, usually methanol with an alkaline catalyst (HAAS et al., 2002).

Oils and fats can be converted into biodiesel by a reaction called transesterification or alcoholysis (Figure 2). In this reaction, triglycerides react with a monoalcohol (methanol or ethanol) in the presence of a catalyst, usually basic, to produce the corresponding esters and glycerol (SERIO et al., 2006). It is important to note that only simple alcohols such as methanol, ethanol, propanol, butanol and amyl alcohol can be used in transesterification (NYE et al., 1983; FREEDMAN et al., 1984). Of these, methanol and ethanol are the most widely used, with the use of methanol in transesterification generally preferred for economic and process-related reasons. In fact, methanol is cheaper than water-free ethanol and has a shorter chain and greater polarity. This latter property makes it easier to separate the esters from the glycerin. However, the use of ethanol can be attractive from an environmental point of view, as this alcohol can be produced from a renewable source and, unlike methanol, does not raise as many toxicity concerns. However, the use of ethanol implies that it is free of water, as well as that the oil used as a raw material has a low water content, otherwise the separation of glycerin will be difficult (FREEDMAN et al., 1984; HATEKEAMA and QUINN, 1994; CONCEIÇÃO et al., 2005).

Oils and fats contain different types of fatty acids linked to glycerol, which, depending on the length of its chain and the degree of unsaturation, is the parameter with the greatest influence on the properties of the vegetable oils and animal fats from which they originate (KNOTHE et al., 2006).

The transesterification of an oil with monoalcohols promotes the breakdown of the triglyceride molecule, generating a mixture of methyl or ethyl esters of the corresponding fatty acids, releasing glycerin as a co-product. The molecular weight of these monoesters is close to that of diesel (RAMOS, 1999). Therefore, transesterification is nothing more than the separation of glycerin from vegetable oil. Around 20% of a vegetable oil molecule is made up of glycerin. Glycerin makes the oil denser and more viscous. During the transesterification process, the glycerin is removed from the vegetable oil, reducing its viscosity (BIODIESELBR, 2006).

Transesterification takes place in a sequence of three consecutive and reversible sub-reactions, with the formation of di- and monoglycerides as reaction intermediates. The stoichiometric proportions are three moles of alcohol per mole of triglyceride, vegetable oil or animal fat. However, some excess alcohol is necessary to increase the conversion yield and allow the subsequent separation of the esters from the glycerol (BIODIESELBR, 2006).

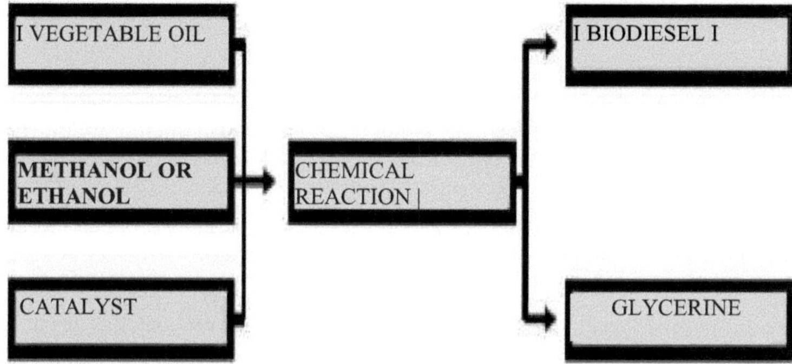

Figure 2: Schematic illustration of the transesterification process.

Source: CEPLAC, 2006.

Once the vegetable oil has been obtained from seeds or kernels, for example (usually by crushing, rolling, cooking and extracting the crude oil), and then purified, it can be converted into biodiesel (Figure 3). In simplified terms, the typical stages in a transesterification process are (BIODIESELBR, 2006):

A) Alcohol and the catalyst are mixed in a tank with an agitator.

B) Vegetable oil is placed in a closed reactor containing the alcohol/catalyst mixture. The reactor is usually heated to approximately

70 C to increase the speed of the reaction, which takes between 1 and 8 hours.

C) At the end of the reaction, when a sufficient level of vegetable oil has been converted, the esters (biodiesel) and glycerin are separated by gravity, and centrifuges can be used to speed up the process.

D) The excess alcohol is separated from the biodiesel and glycerin by evaporation under low pressure (flash evaporation) or by distillation. The recovered alcohol is returned to the process.

E) The biodiesel must be purified and in some cases washed with warm water to remove catalyst residues and soaps.

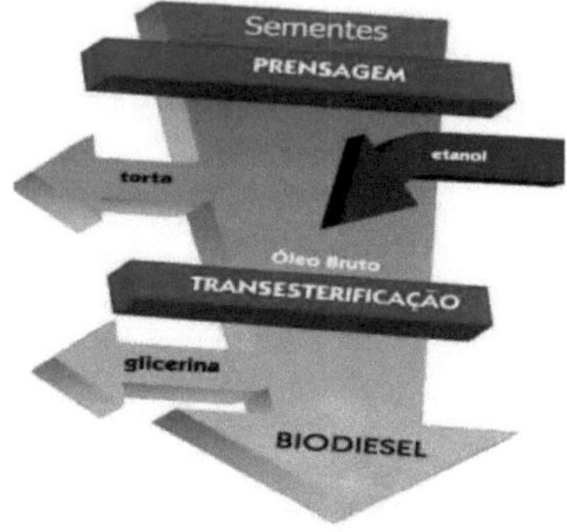

Figure 3: Biodiesel production process.

Source: BIODIESELBR, 2009

2.1.2 Adding biodiesel to diesel in Brazil

Brazil is a major player in the production of fuels from biomass. Since the 1970s, the country has encouraged the development of biofuels for the transportation sector. The national ethanol program (PROALCOOL) was established in 1975 and is today seen as a great success with social, environmental and economic consequences (GOLDEMBERG et al , 2004).

Brazil currently has a new technological and strategic opportunity in the use of biomass: the production of biodiesel. Biodiesel is a fuel derived from vegetable oils (whether new or used) and animal fat (DEMIRBAS and BALAT, 2006).

In Brazil, the first incentive to develop biodiesel production technologies came through the Plan for the Production of Vegetable Oils for Energy Purposes (PROÓLEO), created in 1975 and coordinated by the Ministry of Agriculture. This plan provided for a compulsory blend of 30% in diesel oil until it was completely replaced by biodiesel (CASTELANI, 2008).

In 1980, Brazil was one of the first countries to register a patent for the production of biofuel. However, PROÓLEO was never actually implemented and was replaced by the National Alcohol Program (PROÁLCOOL) (CASTELANI, 2008).

The development of diesel substitutes was tried very hard at the beginning of Proálcool, as a way of further reducing oil consumption and maintaining the production profile of derivatives in line with the capacity of the country's refineries. The process failed for various reasons, including the low prices of diesel at the time, and activities ceased (CASTELANI, 2008).

The government became interested in biodiesel again when its production and consumption began to grow in Europe, especially in Germany; it also saw a way of strengthening family farming and thus improving social inclusion, a very serious problem in Brazil (CASTELANI, 2008).

At the beginning of the century, Ministerial Order 720 of October 30, 2002, established the Brazilian Biodiesel Program (Pró-biodiesel) (BRASIL, 2007), demonstrating the federal government's efforts to move towards sustainable development, in other words, by balancing economic, social and environmental aspects. On December 6, 2004, the National Biodiesel Production Program (PNPB) was officially launched, regulated by Law No. 11,097 of 2005.

The PNPB is an inter-ministerial program in charge of promoting studies in different lines of action (Figure 4) on the feasibility of using vegetable oils for energy purposes, which aims, among other things, to implement sustainable development by promoting social inclusion.

Figure 4: PNPB - National Program for the Production and Use of Biodiesel.

Source: BIODIESEL, 2009

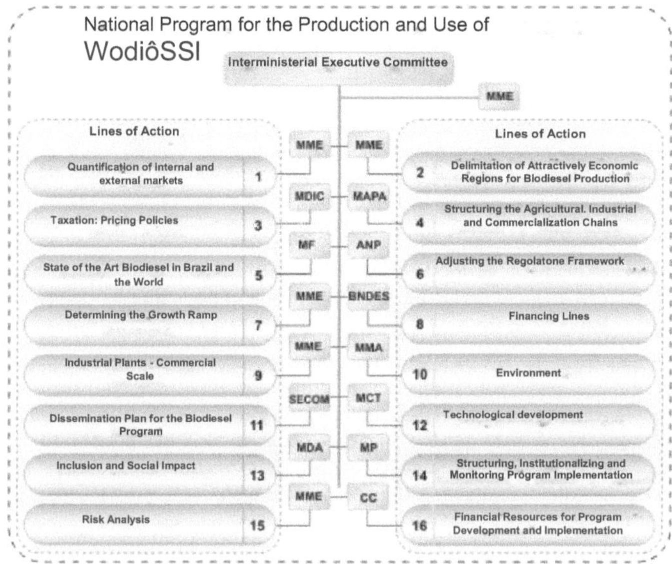

Law 11.097 makes it compulsory to add a percentage of biodiesel to diesel oil sold anywhere in Brazil. From January 2008, a mandatory percentage of 2% (B2) came into force and in 2013 the mandatory percentage will be 5% (B5).

It should be noted that B5 was brought forward to 2010. According to information from the Ministry of Mines and Energy, as of January 1, 2010, the B5 blend became mandatory for all diesel oil consumed in Brazil, except for marine diesel oil. According to current market data, the new blend is expected to generate foreign exchange savings of around US$ 1.4 billion/year due to the reduction in diesel oil imports (MME, 2010).

The PNPB aims to integrate family farmers into the supply of raw materials for biodiesel production, contributing to social equity by generating income. To this end, the Social Fuel Seal was created, granted by the Ministry of Agrarian Development (MDA) to biodiesel producers who promote social inclusion and regional development by creating jobs for farmers who meet the PRONAF criteria. The social seal guarantees mill owners tax benefits, easy access to the best financing conditions and the right to take part in biodiesel auctions, in exchange for providing training and technical assistance to family farmers (CASTELANI, 2008).

2.1.3 Social fuel label

In addition to the economic and environmental advantages, there is the social aspect, which is of fundamental importance, especially considering the possibility of synergistically reconciling all these potentialities of biodiesel (BIODIESEL, 2006).

The planted area needed to meet the B2 blend percentage was estimated at 1.5 million hectares, which is equivalent to 1% of the 150 million hectares planted and available for agriculture in Brazil. This figure does not include areas occupied by pastures and forests. Biodiesel policies allow production from different oilseeds and technological routes, making it possible for agribusiness and family farming to participate (BIODIESEL, 2006).

The cultivation of raw materials and the industrial production of biodiesel, i.e. the biodiesel production chain, has great potential for generating jobs, thus promoting social inclusion, especially when you consider the broad productive potential of family farming. In the Brazilian semi-arid region and the North, social inclusion is even more pressing (BIODIESEL, 2006).

In the semi-arid region, for example, a family's net annual income from growing five hectares of castor beans and an average production of between 700 and 1.2 thousand kilos per hectare can vary between R$2.5 thousand and R$3.5 thousand. In addition, the area can be intercropped with other crops, such as beans and corn (BIODIESEL, 2006).

To further stimulate this process, the Federal Government launched the Social Fuel Seal, a set of specific measures aimed at encouraging the social inclusion of agriculture.

The social framework for projects or companies producing biodiesel gives them access to better financing conditions from the BNDES and other financial institutions, as well as the right to compete in auctions for the purchase of biodiesel. Producing industries are also entitled to exemption from some taxes, as long as they guarantee the purchase of raw materials at pre-established prices, offering security to family farmers. There is also the possibility for family farmers to participate as partners or shareholders in the oil extraction or biodiesel production industries, either directly or through producer associations or cooperatives.

Family farmers also have access to PRONAF credit lines through the banks that operate under this program, as well as access to technical assistance, provided by the companies that hold the Social Fuel Seal, with support from the MDA through public and private partners.

In the 2005-2006 harvest, family farmers who wanted to take part in the biodiesel production chain had an additional PRONAF credit line available for growing oilseeds. As a result, the producer had an additional opportunity to generate income, without leaving the main activity of planting. The aim of this new line was to make the off-season viable, as farmers were able to keep their corn and manioc production, for example, and in the off-season to plant oilseeds. The credit limit and financing conditions followed the same rules as the PRONAF group to which the farmer belonged.

2.1.4 Use of raw materials obtained from family farming

The Social Fuel Seal is an identification component granted by the MDA to biodiesel producers who promote social inclusion and regional development by generating employment and income for family farmers who meet the PRONAF criteria (ASSUMPÇÃO, 2006).

Through this seal, the biodiesel producer will have access to PIS/PASEP and COFINS rates with differentiated reduction coefficients, access to the best financing conditions with the National Bank for Economic and Social Development (BNDES) and its Accredited Financial Institutions, Banco da Amazônia S/A (BASA), Banco do Nordeste do Brasil (BNB), Banco do Brasil S/A (BB), or other financial institutions that have special financing conditions for projects with the social fuel seal. Table 2 illustrates the PIS/PASEP and COFINS rates applied to biodiesel (ASSUMPÇÃO, 2006).

Table 2: PIS/PASEP and COFINS rates applied to biodiesel.

	PIS/Pasep and COFINS (R$/liter of biodiesel)
	Without SealWith Seal fuelsocial fuel social
North, Northeast and semi-arid regions:	
Castor and palm	R$ 0, 150R$0,00
Other raw materials	R$ 0. 218R$0.07
Midwest, Southeast and South Regions	
Any raw material, including castor beans and palm oil	R$ 0, 218 R$ 0,07

Source: ASSUMPTION, 2006.

The biodiesel producer with the social fuel seal assumes the following obligations (ASSUMPÇÃO, 2006):

- Acquire raw materials for biodiesel production from family farmers in a minimum quantity defined by the MDA.
- Enter into contracts with family farmers, negotiated with the participation of a representation of family farmers, specifying commercial conditions that guarantee income and terms compatible with the activity.
- Ensure technical assistance and training for family farmers.

The biodiesel producer is responsible for technical assistance, including its costs, which must be included in the operating costs of biodiesel production. Technical assistance can be provided in-house or outsourced. The MDA, within its means, will support the technical assistance of certified producers in the first few years, through partnerships (ASSUMPÇÃO, 2006).

In order to obtain the seal, the legally constituted biodiesel producer must submit a specific project to the MDA, which will evaluate it in accordance with the norms established in Normative Instruction No. 01 of July 5, 2005. After analysis and auditing, the MDA will publish an extract in the Federal Official Gazette which will grant the biodiesel producer access to the benefits of the seal (ASSUMPÇÃO, 2006).

In the Northeast and semi-arid regions, biodiesel producers will have to buy at least 50% of their raw materials from family farms. In the Southeast and South, this minimum percentage is 30% and in the North 10% (ASSUMPÇÃO, 2006).

2.1.5 Marketing biodiesel

SRF Normative Instruction No. 516 of 22/02/2005 provides for the Special Registration of biodiesel producers or importers and SRF Normative Instruction No. 526 of 15/03/2005 (art. 52 of Law No. 10.833 of 29/12/2003 and art. 4 of Provisional Measure No. 227 of 06/12/2004) provides for the levying of PIS/PASEP and COFINS contributions on revenues from the sale of biodiesel.

In addition to the provisional measure, Decree No. 5.297 of December 6, 2004 was issued, which sets out the coefficients for reducing the PIS/PASEP and COFINS rates levied on the production and sale of biodiesel, and the terms and conditions for using differentiated rates, i.e. this decree highlights aspects relating to tax benefits (ASSUMPÇÃO, 2006).

The ANP's resolutions, among other things, establish the quality control procedures for the production of biodiesel and diesel/biodiesel blends; the rules for the installation and tankage of the Retail Resale Transporter (TRR), as well as the product quality analysis that the TRR must carry out when it receives the product; the regulations for importing diesel oil and biodiesel; how the resale station should proceed as to whether or not to display the distributor's brand on the product it resells; and the rules for exporting biodiesel (ASSUMPÇÃO 2006).

Biodiesel production can only be carried out by legal entities set up as a limited or joint-stock company, with headquarters and administration in the country, and which have authorization from the ANP, maintain a Special Registration with the Federal Revenue Office of the Ministry of Finance and have subscribed and paid-in capital of R$ 500,000.00 (CASTELANI, 2008).

ANP authorization requires, among other documents, an environmental license, an operating permit, a Fire Department Inspection Report and a technical report containing information on the biodiesel plant's production process and capacity.

According to ANP Resolution 41, the biodiesel producer can only sell the product: (a) to the refinery authorized by the ANP; (b) to the exporter authorized by the ANP, (c) directly to the foreign market, when it is authorized by the ANP to export biodiesel, or (d) to the distributor of petroleum-derived liquid fuels, fuel alcohol, biodiesel, diesel/biodiesel blends specified or authorized by the ANP and other automotive fuels (CASTELANI, 2008).

As such, biodiesel producers cannot sell directly to final consumers (farmers, transporters and others). Direct sales to final consumers are only allowed in cases of experimental use, duly authorized by the ANP, under the terms of Ordinance 240 of August 25, 2003. Unspecified products are those whose characteristics have not been defined by the ANP and which are used in a mixture with hydrocarbons derived from oil, natural gas or alcohol, or as a substitute for them in processes or equipment, as is the case with B100 (100% biodiesel) (CASTELANI, 2008).

2.1.6 Advantages of incorporating biodiesel into the Brazilian energy matrix

There are several potential sources of oilseeds in Brazil for biodiesel production, given the wide diversity of our ecosystem. This is a comparative advantage that the country has over all other oilseed producers (LADETEL, 2005).

In addition to these advantages, there are others that are more specific to the use of biodiesel (LADETEL, 2005):

a) Ecological advantages: The emission of combustion gases from engines running on biodiesel does not contain sulphur oxides, the main cause of acid rain and respiratory tract irritation. The agricultural production that produces the raw materials for biodiesel captures CO2 from the atmosphere during the growing season, and only part of this CO2 is released during the combustion process in the engines, helping to control the "greenhouse effect", which causes global warming.

b) Macroeconomic advantages: The expansion in demand for agricultural products should generate employment and income opportunities for the rural population; biodiesel production could be carried out in locations close to where the fuel is used, avoiding the unnecessary cost of redundant movement; the domestic use of vegetable oils will make it possible to circumvent the low prices that prevail in world markets demeaned by protectionist practices.

c) Diversification of the energy matrix through the introduction of biofuels. A specific methodology needs to be defined for studies of investment alternatives for the introduction of new technologies for the production and distribution and logistics of biofuels.

d) Financial advantages: Biodiesel production will make it possible to meet the targets proposed by the Kyoto Protocol, the Kyoto Protocol's

   Clean Development, enabling the country to participate in the "carbon bonus" market.

e) Regional development: The dynamic of globalization is continuous renewal, and it is a reality that every capitalist consumption pattern is dictated by the highest scales, in other words, by the countries with the most advanced technological standards. It is therefore vital to restructure the production system, demonstrating the need for productive innovations, including the creation of a competitive biodiesel chain as a local development response to the global challenge.

2.1.7 Obstacles to the use of biodiesel in Brazil

Some sectors of social movements and environmentalists are harsh critics of the new technology (PINTO and MENDONÇA, 2007). They point to alarming data such as the possible increase in deforestation, the expansion of monocultures and all the resulting problems such as the loss of biodiversity, the damage to food sovereignty, the rise in pollution rates caused by the increased use of chemical inputs in crops and the greater vulnerability of small producers.

Data from 2007 showed that, three years into the PNBP, soy accounted for 55% of the raw material used to produce biodiesel in Brazil, castor beans accounted for 20% and the rest was divided between other oilseeds such as turnip rape and palm oil (OLIVEIRA, 2007). Today, soy accounts for 90%. Soybean production has been expanded into the cerrado region, in disregard for biodiversity, cultivated in large areas of monoculture, stimulating concentrations of land and income, and its production system is highly mechanized, which restricts the social inclusion of small farmers.

Souza (2004) evaluated the employment potential of some oilseeds and the land occupied by family farming. He found that soybean production uses 20 hectares of land to employ one family, while the same family would occupy 16 hectares of peanuts (mechanized tillage). Growing babassu and oil palm would require 5 hectares per family and castor beans 2 hectares per family. It can therefore be seen that the production of biodiesel from soya is at odds with the context in which the PNPB was created and has hindered its convergence towards social inclusion.

Table 3 summarizes the advantages and disadvantages of incorporating biodiesel into the national energy matrix.

Table 3: Summary of the advantages and disadvantages of biodiesel

ADVANTAGES	DISADVANTAGES
Ecological	Possible increase in deforestation
Macroeconomic	Monoculture Expansion
Diversification of the energy matrix	Damage to food sovereignty
Financial	The increase in pollution caused by the use of chemical inputs.

Source: LADETEL, 2005.

2.1.8 Economic and social aspects of biodiesel

Biodiesel is generally more expensive to produce than oil, requiring government subsidies since its production is not yet an economically viable option (WASSELL JR. & DITTMER, 2007). The cost of producing biodiesel in Europe and the USA is 50% higher than that of mineral diesel without taxes (OECD, 2006 apud PRATES et al., 2007).

Prates and collaborators (2007) mention that a directive was launched in Europe in 2003, which authorizes tax or partial tax exemption on biofuels. However, the use of biofuels is justified by positive externalities, such as the environmental issue - discussed above - and the promotion of agribusiness (PRATES et al., 2007).

With regard to the social aspect, it is worth noting that biodiesel production generates jobs and income for rural workers, "being a vector for the internalization of development, the reduction of regional disparities and the settlement of populations in their habitat, especially by adding value to the production chain and integrating different dimensions of agribusiness" (PLANO NACIONAL DE AGROENERGIA, 2006 p. 16).

In addition, the use of biodiesel leads to a reduction in pollutant emissions, thus representing a significant improvement for public health, especially in large cities (BOROMI, 2004 apud PENTEADO, 2005).

2.1.9 Main raw materials for biodiesel production and the focus on family farming

The Biodiesel Program is part of the Brazilian government's policy to promote the production of alternative fuels derived from vegetable oils. The main raw materials for the national production of biofuel are: soy, corn, sunflower, peanuts, cotton, canola, castor, babassu, palm and macauba, among other oilseeds found in the country. Fuel can also be obtained from waste oils (from industrial kitchens) and animal fats (beef tallow, fish oils, lard, among others).

Although the country has a great diversity of agricultural inputs for the production of vegetable oils and, consequently, biodiesel, many crops are still extractive in nature and there are no commercial plantations to assess their real potential. In this respect, soy, which accounts for 90% of Brazilian vegetable oil production, palm oil, babassu, sunflower and castor beans are the main options (MELLO et al, 2007).

Table 4 shows some characteristics of oilseed crops with potential for use for energy purposes. According to Holanda (2004), among the permanent crops, oil palm and babassu stand out. The oil palm crop can be an important source of vegetable oil, as it has an extraordinary productivity of around 6,000 kg of oil per ha/year, a figure around 25 times higher than that of soybeans. However, this figure is only reached 5 years after planting.

Table 4: Characteristics of oilseed crops with potential for use for energy purposes.

Species	Origin of the oil	Oil content (%)	Harvest month/year	Yield (tons of oil/ha)
Oil palm	Pulp	22	12	3,0 - 6,0
Coconut	Fruit	55 - 60	12	1,3 - 1,9
Babassu	Almond	66	12	0,1 - 0,3
Sunflower	Grain	38 - 48	3	0,5 - 1,9
Rapeseed / Canola	Grain	40 - 48	3	0,5 - 0,9
Castor beans	Grain	45 - 50	3	0,5 - 0,9
Peanuts	Grain	40 - 43	3	0,6 - 0,8
Soy	Grain	18	3	0,2 - 0,4
Cotton	Grain	15	3	0,1 - 0,2

MAPA, 2005.

With regard to babassu, its kernel contains an average of 66% oil, but the kernel only represents 7% of the fruit's weight. It has only 4% oil in total and oil production per hectare is low. However, the existence of 17 million hectares of forest, where the babassu palm predominates, and the possibility of making full use of the coconut (oil, charcoal, etc.) make it possible to use it in an economically viable way and with a high degree of social insertion (CHIARANDA; ANDRADE JÚNIOR; OLIVEIRA, 2005).

The Ministry of Agriculture has assessed the development of production chains for different vegetable oils in each state and region of the country. Thus, biodiesel production must respect the specificity of each region by producing what, in a way, will give it a greater comparative advantage (MELLO et al., 2007).

Figure 5 shows the map with the regions and their respective crops for biofuel production. It should be noted that in the northeast region, which includes the state of Maranhão, the subject of this study, the main raw materials for biodiesel production are: babassu, soy, castor beans, palm and cotton. However, in the work carried out by the Ministry of Agriculture, the potential of each state for the production of these raw materials is not presented, nor is the geographical location of the production areas.

Studies carried out by the Ministry of Agrarian Development, the Ministry of Agriculture, Livestock and Supply, the Ministry of National Integration and the Ministry of Cities show that for every 1% of family farming's participation in the country's biodiesel market, based on the use of B5, it would be possible to generate around 45,000 jobs in the countryside, at an average cost of R$4,900.00 per job (HOLANDA, 2004).

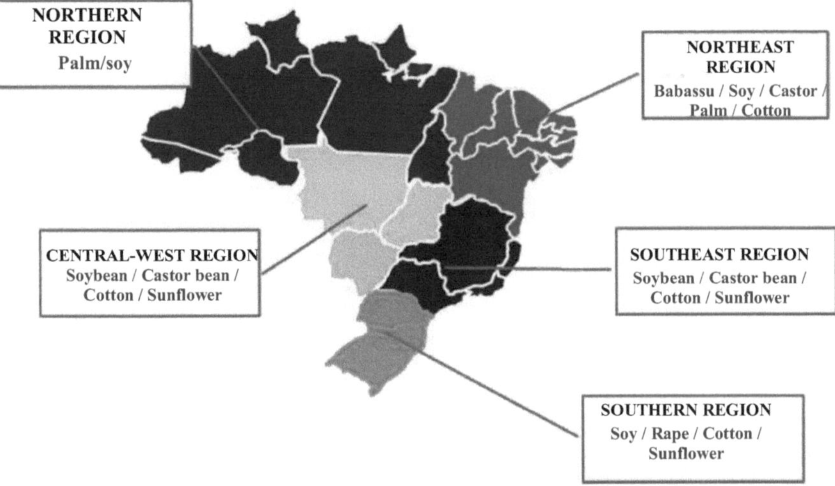

Figure 5: Map of Brazilian regions and their crops for biodiesel production.
Source: Embrapa apud Chiaranda, Andrade Júnior, Oliveira (2005).

The same study points out that every R$1.00 invested in family farming generates an additional R$2.13 in annual gross income, which means that family income would double with participation in the biodiesel market (HOLANDA, 2004).

These estimates justify the social connotation given to the biodiesel program, but it should be noted that the viability of the program depends on the raw materials that will be used to produce the biofuel.

2.2 GLYCERINE

Glycerin is the name of the commercial product that consists of glycerol and a small amount of water. Glycerol is an alcohol (C3H8O3) named 1,2,3 propanetriol. Its chemical structure is shown in Figure 6 (BAILEY and HUI, 2005).

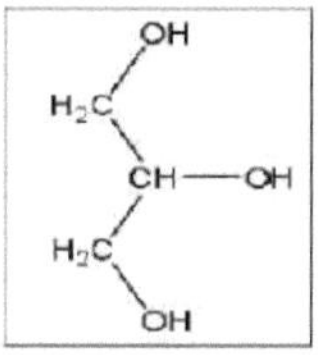

Figure 6 - Molecular structure of glycerol.

Glycerol is obtained as a by-product in the conversion of fats and oils to fatty acids or fatty acid methyl esters. This type of glycerol is known as native or natural glycerol, in contrast to the synthetic glycerol obtained from propene. Other production methods, such as sugar fermentation or carbohydrate hydrogenation, are not of industrial importance (ULLMANN'S et al, 1988).

Glycerol was discovered in 1779 by Scheele through the saponification of olive oil. In 1813 Chevreul showed that fats are fatty acid esters of glycerol (COSTENARO, 2009).

The first industrial use of glycerol was reported in 1866, when Nobel produced dynamite, in which glycerol trinitrate (nitroglycerin) was stabilized by adsorption on diatomaceous earth (COSTENARO, 2009).

The most important synthesis of glycerol, using propene as a starting material, was developed at the end of the 1930s by I.G. Farben in Germany and Shell in the United States (ULLMANN'S et al, 1988).

Natural glycerin is essentially a by-product of the production of certain animal and vegetable fat processes (BAILEY; HUI, 2005).

The main co-product of biodiesel production is glycerin (10% of production), in its raw state, containing some impurities that must be eliminated for the production of other products, such as resins, soaps and soap (COSTENARO, 2009).

Glycerin is one of the most versatile and valuable chemical substances for humans, as it has a unique combination of physical and chemical properties that are used in many products. Glycerin has more than 1500 known uses, such as in cosmetics, food products, medicines and personal care. Furthermore, it is highly stable under typical storage conditions and is compatible with various other chemical materials (BONNARDEAUX, 2006).

The crude or "blond" glycerin obtained in the transesterification process is around 80% glycerol. It contains water, free fatty acids, residual inorganic salts from the catalyst, unreacted mono- and di-glycerides, methyl esters and other organic compounds in smaller proportions. In this condition, glycerin is yellow to brown in color. For this reason, crude glycerin has few uses and low commercial value. Its main uses, however, require it to be purified.

Typically, the first step in the glycerin purification process is to separate the oil from the soaps and other organic impurities by filtration and/or centrifugation. The final purification is usually done under vacuum distillation, followed by bleaching with activated carbon for large-scale operations. Other processes have been developed to reduce purification costs (EET, 2008).

The scarcity of studies on alternative and economical methods for purifying crude glycerin from biodiesel production indicates the need to develop purification routes depending on the type of transesterification, the physical parameters and the quality of the crude glycerin generated (SALVADOR; MACHADO; SANTOS, 2006). Or, the development of new industrial applications for this type of product.

2.2.1 Glycerin supply and demand

In Brazil, the installed production capacity in 2006 was almost 32,000 tons/year, with Unilever accounting for just over 50% of this production capacity, out of the 11 manufacturing companies. Reported production in the same year was approximately 14,000 tons, almost all of which was sold on the domestic market. The main consumer of glycerin today is the cosmetics industry (65.1%), followed by the paint and varnish industry, as shown in Figure 7 (ABIQUIM, 2008).

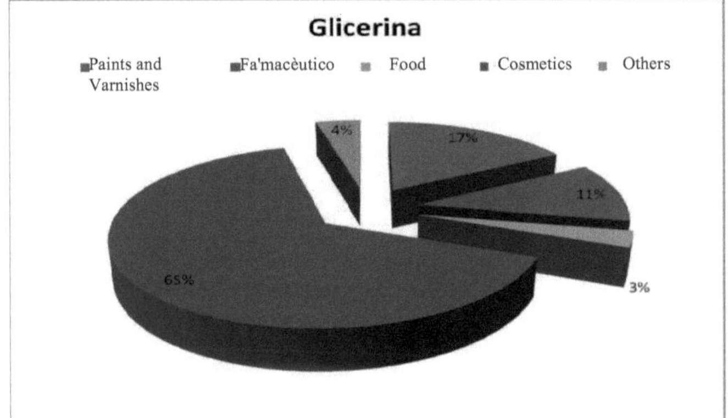

Figure 7: Main destinations for glycerin sold on the domestic market.

Source: ABIQUIM, 2008.

In the cosmetics industry, glycerin is used in toothpastes, moisturizers, aftershave lotions, deodorants, lipsticks and make-up. In the pharmaceutical industry, glycerin is used in capsules, suppositories, anesthetics, Using data provided by the Ministry of Development, Industry and Foreign Trade, Figure 8 was constructed, showing the quantities of glycerin imported and exported between 1998 and 2007. It can be seen that between 1999 and 2000, the country exported around 600 tons a year. Between 2001 and 2003 there was a period of self-sufficiency. In 2004, the country had to import almost 1000 tons to meet its needs. From 2007 onwards, exports reached volumes of over 5000 tons per year.

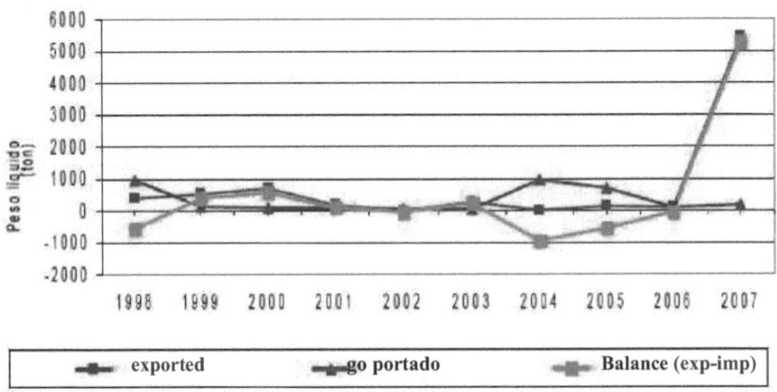

Figure 8: Glycerin import and export volumes between 1998 and 2007.
Source: MDIC, 2008.

113 kg of crude glycerin are generated for every ton of biodiesel produced. Considering the 2006 scenario, the introduction of B2 implies the co-production of around 80,000 tons of crude glycerin per year.

As in 2006 practically all the glycerin produced by the current soap and detergent factories was consumed on the domestic market and the trade balance between the quantities exported and imported is practically zero, the current scenario does not show any possibility of absorbing the 80,000 tons of glycerin on the domestic market generated in the transesterification process to supply B2. In 2007, the volume of glycerin exported almost quintupled, as it is being co-generated in the manufacture of biodiesel and needs to be disposed of.

Another important observation is that the price paid for the exported product has fallen significantly in recent years. Between 1998 and 2001, the average price was US$ 1.05/kg. Between 2005 and 2007, the average export price was US$ 0.36 / kg (MDIC, 2008).

The worldwide overproduction of glycerine created as a co-product of the biodiesel production process is leading traditional glycerine plants to close and other plants that use it as a raw material to open. The Solvay group has built a new epichlorohydrin plant in Tavaux, France, which uses glycerin instead of propylene glycol as a raw material. The Belgian company buys glycerine from biodiesel producers at competitive prices.

Epichlorohydrin is used to make epoxy resins, reinforcing agents for paper and other products. The Solvay group says that this new process called Epicerol was made possible by the creation of a new class of catalysts covered by 11 application patents. The giant agro-industrial group Archer Daniels in the US has inaugurated plants that produce propylene glycol from glycerin instead of propylene oxide, in a process that also employs advanced catalysis. Companies that have traditionally produced glycerin are closing down. Dow Química closed its plant in Texas due to the flood of glycerine on the market produced in the bioidesel process. Procter & Gamble Chemicals has closed its natural glycerine refinery in England.

In Brazil, the company Nova Petroquímica has announced investments of US$ 50 million to build an industrial plant that will produce the so-called "green polypropylene", i.e. from the glycerin generated as a by-product of biodiesel production (JORNAL DE PLÁSTICOS, 2008).

2.2.2 Actions by the Brazilian biodiesel production and use network to use crude or rectified glycerin

In Brazil, 100,000 tons of glycerin were produced in 2008; by 2010, 250,000 tons are expected for an annual consumption in the Brazilian market of 30,000 tons/year (BIODIESELBR, 2009b). Therefore, there is a real need to develop new technological routes for using the glycerin generated in biodiesel production. The chemical industry, which uses glycerin as a raw material in various processes, is a very attractive target for absorbing this surplus, thus justifying the allocation of resources for research in the areas of fine chemicals, composites and energy generation.

About two years ago, a thematic network was established within the National Program for the Production and Use of Biodiesel to develop applied research into the use of co-products from the biodiesel production chain, namely cake and/or bran and glycerin. Estes foram os projetos de pesquisa selecionados para financiamento e que hoje se encontram em desenvolvimento: (a) produção de aditivos oxigenados para gasolina a partir da glicerina oriunda da produção do biodiesel - Dr. Claudio José de Araújo Mota, UFRJ (RJ); (b) estudo e otimização de processos de obtenção de éteres a partir do glicerol subproduto obtido do processo de transesterificação de óleos vegetais - Prof. Dr. Fernando Carvalho Silva, UFMA (MA); (c) development of additives for ultra-deepwater oil well drilling fluids from glycerin - Dr. Regina Sandra Veiga Nascimento, UFRJ (RJ); (d) obtaining polyols derived from glycerin as a raw material for the production of polyurethanes - Dr. Cesar Liberato Petzhold, UFRGS (RS); (e) use of glycerol in the production of polymeric membranes for fuel cells - Dr. Maria A. F. César- Oliveira, UFPR (PR); (f) manufacture of composites derived from starch plasticized with crude glycerin and reinforced with natural fibres -

Dr. Fernando Wypych, UFPR (PR); (g) adsorptive processing of by-products from biodiesel production - Dr. Cesar Augusto Moraes de Azevedo, UFRGS (RS). Cesar Augusto Moraes de Abreu, UFPE (PE); (h) development of plasticizers for PVC from glycerin and its derivatives - Dr. Sonia Faria Zawadzki, UFPR (PR); (i) use of glycerol in nanostructured bioplastics - Dr. Sonia Faria Zawadzki, UFPR (PR).

Cristina Tristão de Andrade, IMA-UFRJ (RJ); (j) gasification of biodiesel production waste for energy self-sufficiency, Dr. Sérgio Peres Ramos da Silva, UPE (PE); (k) energy use of glycerin, Dr. Donato A. Gomes Aranda, UFRJ (RJ); (l) energy use of biogas from glycerin (anaerobic route), Dr. Maria de Los Angeles Perez Fernandez Palha, UFPE (PE); (m) simulation of advanced oil recovery in a reservoir micro-cell by injecting crude glycerin (GB), Dr. Cristina Maria A. L. T. da Mata Hermida Quintella, UFBA (BA); and (n) catalytic oxidation of glycerin applied to the production of biodegradable detergents, Dr. Nelson Medeiros de Lima Filho, UFPE (PE).

As you can see, a number of research and development opportunities are being evaluated and they will certainly lead to technically and economically viable ways of using glycerin. One of them, whose main development center is the Department of Mechanical Engineering at the Federal University of Uberlândia, deals with the production of synthesis gases from different types of glycerin water, which will certainly open up excellent prospects for using this material in second-generation fuel production units.

2.3 DESCRIPTION AND CHARACTERISTICS OF COOPERATIVES

At the beginning of the 20th century, some cooperatives began to appear, inspired by models brought by foreign immigrants or by a few idealistic Brazilians, who were aware of the success of cooperative credit associations for small farmers in Germany and Italy. Rio Grande do Sul was at the forefront of cooperative credit. "Subsequently, cooperatives in various branches multiplied throughout the country" (PINHO, 2004, p.14). As Queiroz (1998, p.12) states, "cooperatives are growing all over the world in the search for professional and social improvements for workers".

According to Brazilian legislation (art. 4 of Law No. 5764, of January 16, 1971), cooperatives "are partnerships, with their own legal form and nature, of a civil nature, not subject to bankruptcy, set up to provide services to members" (BRASIL, 1998).

According to the Organization of Cooperatives of Brazil (OCB, 2009), "a cooperative is an autonomous association of people who voluntarily unite to satisfy common economic, social and cultural aspirations and needs, through a collectively owned and democratically managed enterprise."

They are based on values of "mutual aid and responsibility, democracy, equality, fairness and solidarity. In the tradition of their founders, cooperative members believe in the ethical values of honesty, transparency, social responsibility and concern for their fellow man" (OCB, 2009).

Cooperatives are guided by certain principles, through which they put their values into practice (OCB, 2009):

- Voluntary and free membership.
- Free and democratic management.
- Members' economic participation.
- Autonomy and independence.
- Education, training and information.
- Intercooperation.

- Interest in the community.

This last principle - interest in the community - means that cooperatives work for the sustainable development of their communities through policies approved by the members. According to Cançado and Gontijo (2004), there are two aspects to the definition given to this principle: "Firstly, cooperatives, as people's organizations, tend to be closely linked to the community where the members live, and in this way, the development of this community is directly reflected in the members. The second aspect concerns the way in which the cooperative acts in the community, in other words, the very definition of its action policies, which must
be approved by its members, so this principle must be applied in conjunction with that of Democratic Management".

"One difficulty in applying this principle in practice is the growing scarcity of resources to manage these actions, given the tendency for margins to shrink and the consequent reduction in results, which is not only true of cooperative organizations." This requires "great creativity and flexibility to find solutions that combine scarce resources with satisfactory results" (BRAGA et al. apud CANÇADO E GONTIJO, 2004, p.14).

According to Pinho (2004, p.121), in cooperative societies there is "equality of rights and obligations of the members based, above all, on the rule established at general meetings that each member has the right to one vote [...] regardless of their participation in the share capital".

According to Perius (1983), members are free to join cooperative organizations, as long as they make the decision to cooperate, as long as this brings benefits, which can be summarized as maximizing the remuneration for professional activities. The cooperative legal relationship is institutional in nature, as it is based on submission to previously established statutory rules. However, in recent years, most cooperatives have established requirements for members to join, such as specifying the product to be received, quantity and quality.

Cooperatives can be structured in different ways: functional structuring, structuring represented by strategic business units and process structuring. Functional structuring is the most common, consisting of: a general assembly as the highest decision-making body; a supervisory board to oversee the cooperative's budget execution; a board of directors with the task of managing the cooperative enterprise; a general board; and financial, production, marketing and human resources management (OLIVEIRA, 2001).

According to Pinho (2004, p.118), historically, the agricultural cooperative "is the best structured Brazilian cooperative branch".

"As a result of globalization, the growing international competition and agricultural protectionism in developed countries, agricultural cooperatives have been forced to look for solutions to guarantee the safety of their products.
competitiveness and sustainability of agriculture, of the cooperative itself as a business and of the cooperative body" (PINHO, 2004, p.118).

In the same vein, Bialoskorski Neto (2005) points out that due to the intense transformation that agricultural activity has undergone in recent years, what used to be a subsistence and self-sufficient activity has become a unit dependent on the market and the input and processing industries. This has forced agricultural organizations to adapt to market demands and seek alternatives, especially small organizations, in order to survive in a highly competitive market.

In this sense, an alternative would be the formation of cooperatives. For developing countries, the formation of cooperatives generates dynamism from an economic and social point of view, and their contribution to the development of agricultural activities stands out. In Brazil, the growth of cooperatives is fundamentally due to this sector (NASCIMENTO, 2000).

Cooperativism develops intensively in the primary sector of the economy due to the market structures found there. The reason for this is that agriculture is characterized by interacting with highly concentrated markets, as is the case with the basic inputs needed, the processing and distribution of production, according to the chain perspective.

Thus, according to Bialoskorski Neto (2005, p. 236), "cooperative business economies are situated between the private economies of the cooperative members, on the one hand, and the market, on the other, appearing as intermediate structures, formed from spontaneous collective action".

Therefore, an important reason for the existence of this type of organization is that they make it possible to reduce risks and add value for rural producers who, in many cases, would not be able to relate to concentrated markets on their own.

From the above, it is possible to understand the importance of cooperatives for agribusiness, especially with regard to small producers who, if they don't unite in larger corporations, will find it difficult to survive in an increasingly globalized and competitive environment. It can be seen that cooperatives arose with the aim of solving a social problem of the exclusion of small artisans. Therefore, they seek a balance between the social and economic spheres, having to deal with the dilemma: the values and needs of the producers and the people.
cooperative members versus competition and market values. As explained by Begnis and collaborators (2004), "the primary purpose of a traditional cooperative is the economic and social development of its members". However, in addition to the social issue, it is important to include the environmental issue in order to achieve sustainability.

2.4 FUNDING SOURCES

The sources of funding for the Biodiesel Program can be internal or external. The main financing agent for the Brazilian Biodiesel program is the National Bank for Economic and Social Development (BNDES), whose main operating conditions for the program are (LIMA, 2004):

- Raising the maximum percentages of participation in total investment and total terms.
- Granting a guarantee for external financing.
- Relaxation of the guarantee rules in force in 2004.
- Simplification of the analysis and contracting processes.

Another financing agent is Banco do Brasil, which has BB BIODIESEL, the BB Program to Support the Production and Use of Biodiesel (ASSUMPÇÃO, 2006):

- The program aims to support the production, marketing and use of biodiesel as a source of renewable energy and as an activity that generates employment and income.
- Assistance to the production sector is provided through financing lines for costing, investment and marketing, helping to expand biodiesel processing in the country by encouraging the production of raw materials, the installation of agro-industrial plants and marketing.
- The program systematically benefits the various components of the biodiesel production chain:
    a) In agricultural production: with credit lines for funding, investment and commercialization, available to finance family and business rural producers.
    b) In industrialization: BNDES Biodiesel, PRONAF Agroindustry, PRODECOOP, Agroindustrial Credit (acquisition of raw materials), in addition to the lines available for the industrial sector.

The main criterion considered by the Bank when granting credit, in addition to the specific requirements of each line, is the guarantee of commercialization of both agricultural production and biodiesel. Initially, priority is given to oil palm, castor bean, soybean, cotton (seed), sunflower and turnip rape crops, taking into account agricultural zoning and regional suitability (BANCO DO BRASIL, 2006b).

Lima (2004) points out that "in the Kyoto Protocol's Clean Development Mechanism (CDM), part of the commitment to reduce greenhouse gas emissions by developed countries can be carried out in developing countries".

In July 1999, the World Bank created the Prototype Carbon Fond (PCF), a fund to finance projects aimed at mitigating the effects of climate change and promoting sustainable development, with resources of around 150 million dollars. To make up this fund, governments and companies from developed countries contribute resources and technology for the projects. The PCF passes on these resources to finance projects in developing countries (LIMA, 2004).

Another external source is the Inter-American Development Bank (IDB) which, according to Lima (2004), granted Colombia a loan of US$10 million to promote energy efficiency.

When financing with funds from abroad, it is necessary to take into account the problem of exchange rate risk, finding an alternative of how to make this financing available without leaving the exchange rate risk to the borrower (ASSUMPÇÃO, 2006).

2.4.1 National Program to Strengthen Family Farming - PRONAF Investment

PRONAF aims to provide financial support for agricultural and non-agricultural activities carried out using the direct labor of the rural producer and his family. Non-agricultural activities include services related to rural tourism, craft production, family agribusiness and other rural services that are compatible with the nature of the rural holding and the best use of family labor (BNDES, 2009).

The program covers the entire national territory. The beneficiaries of PRONAF are all those who (BNDES, 2009):

- Explore a plot of land as owner, squatter, tenant, partner or concessionaire of the National Agrarian Reform Program.
- They live on the property or nearby.
- Do not have, in any capacity, an area greater than 4 (four) fiscal modules, quantified according to the legislation in force.
- Obtain at least 70% of their family's income from farming and non-agricultural activities.
- Have family work as the predominant form of exploitation of the establishment, only occasionally using salaried labor, according to the seasonal requirements of the agricultural activity, and may keep up to 2 (two) permanent employees.
- Have obtained gross annual family income in the last 12 months preceding the application for the Declaration of Aptitude to PRONAF - DAP above R$ 6,000.00 and up to R$ 110,000.00, including income from activities carried out on and off the farm by any member of the family, excluding social benefits and social security income from rural activities. Credits can be granted individually or collectively.

The interest rate varies from 1 to 5% depending on the framework, and the payment period is up to 12 years, with a grace period of up to five years (BNDES, 2009).

2.4.2 Cooperative Development Program for Adding Value to Agricultural Production - PRODECOOP

The aim of PRODECOOP is to increase the competitiveness of the agro-industrial complex of Brazilian cooperatives by modernizing production and marketing systems (BNDES, 2009a).

The items that can be financed are (BNDES, 2009a):

- Studies, projects and technology.
- Civil works, installations and other fixed investments.
- New domestic machinery and equipment accredited by the BNDES and inherent to the cooperative's production/processing.
- Pre-operating expenses.
- Import expenses, in national currency, linked to the import of equipment.

- Working capital associated with the investment project, subject to the BNDES Automatic limits.
- Working capital not associated with investment projects.
- Training.
- Payment of shares linked to the project to be financed.
- New domestic machinery and equipment accredited with the BNDES, also on an isolated basis, when destined for modernization within the scope of the sectors and actions supported by the Program.

The sectors that can be financed are (BNDES 2009a):

- Industrialization of oilseed derivatives.
- Relocation of oilseed processing plants.
- Implementation of energy generation and co-generation systems and connection lines, for own consumption, as part of an agro-industry project.
- Installation, expansion and modernization of industrial units for the production of alcohol, sugar and biodiesel.

The interest rate is 6.75% per year, including the accredited financial institution's remuneration of 3% per year. The participation level is up to 90%.

During the period from July 1, 2009 to June 30, 2010, each cooperative can contract financing of up to R$50 million for projects in a single state, in one or more operations. The limit for financing working capital not associated with investment projects is up to R$20 million, to be deducted from the credit limit (BNDES, 2009a).

The limit of R$ 50 million may have the BNDES' participation increased by up to 100% when the additional resources are earmarked for the cooperative's own ventures in other Federation Unit(s) or for ventures carried out within the scope of a central cooperative (BNDES, 2009a).

Chapter 3

3. METHODOLOGY

The research carried out in this study is descriptive because it exposes the characteristics of a certain population or phenomenon, and exploratory because it was carried out in an area in which there is little accumulated and systematized knowledge, as well as interventionist because it aims to interfere in the reality studied in order to change it (VERGARA, 2000, p. 46-49).

In terms of the means used, this is a field, documentary and bibliographical study. Field, because it is an empirical investigation carried out in the place where a phenomenon occurred and has elements to explain it; documentary, because it is carried out using documents kept in public and/or private bodies and bibliographical, because it uses material accessible to the general public, such as books, magazines, newspapers, the Internet, etc. (VERGARA, 2000, p. 46-49 and GIL, 1991, p.47-52).

The research approach is "dialectical", because in order to understand the object of the research it was necessary to analyze the links, mediations and contradictions. It is an analysis that goes beyond the technical issue and addresses the social and economic problem (TRIVINOS, 1987, p. 49).

Therefore, based on the literature on biodiesel, a study was carried out mentioning its definition, characteristics, production technology through the transesterification process, the use of raw materials from family farming, commercialization, its advantages and disadvantages in terms of inclusion in the energy matrix and its social economic aspect.

In order to achieve the objectives set out in this dissertation, a socio-economic analysis of the state of Maranhão was carried out (item 4) to select and evaluate the availability and location of the most promising raw material for biodiesel production in Maranhão; to calculate the estimated production of biodiesel from the selected raw material; and to size the plant to meet local raw material production. The available workforce and the existence of cooperatives in the selected region were also analyzed. Next, an economic analysis of biodiesel production was carried out (item 5). To this end, quotations for plants were made with manufacturers, the necessary infrastructure was surveyed and the co-products formed in the production chain were surveyed. Finally, the socio-economic impact for the cooperative's members of setting up a biodiesel plant in the region was estimated. In addition, based on the fixed and variable costs, financial income and payback, within the sales forecast for babassu biodiesel and its by-products, the project was analyzed and simulated to fit into the PRONAF credit line.

As a basis for consulting and analyzing data, we used information published by government agencies in the state of Maranhão, such as IMESC - Maranhão Institute for Socio-Economic and Cartographic Studies by the federal government, IBGE - Brazilian Institute of Geography and Statistics, EMBRAPA - Brazilian Agricultural Research Corporation, SEPLAN - State Secretariat for Planning and Economic Development, EPE - Energy Research Corporation and BNDES - National Bank for Social Development, as well as national and international technical and scientific magazines/journals.

Chapter 4

4. ANALYSIS CARRIED OUT TO ACHIEVE THE OBJECTIVES SET OUT IN THIS RESEARCH

As mentioned in the introduction (item 1), in order to achieve the proposed objectives, the research carried out in this dissertation was divided into two phases. The first was a socio-economic analysis of the state of Maranhão (item 4.1) and the second was an economic analysis of biodiesel production based on the potential raw material and the location of the plant based on the analysis carried out in the first phase (item 4.2).

4.1 SOCIO-ECONOMIC ANALYSIS OF THE STATE OF MARANHÃO

According to the Ministry of Foreign Affairs, the state of Maranhão, located on the northern coast of Brazil, occupies an area equivalent to 333,365.6 km^2 , bordered to the north by the Atlantic Ocean, with a coastline of 640 km. To the east it is bordered by the state of Piauí, to the south and southwest by the state of Tocantins and to the west by the state of Pará. The predominant climate in the state is tropical and its relief has two distinct regions, which include the coastal plain and the tabular plateau. The coastal plain is made up of marshy lowlands, tablelands and extensive beaches. Noteworthy are the large expanses of sand dunes and the jagged coastline on some stretches of the coast, especially where the bays of São Marcos and São José are formed. The other regions are made up of plateaus, which form slopes with escarpments, known as serras. In the northwestern part of the state lies the so-called Maranhão Amazon, which is characterized by forest vegetation and an equatorial climate (MRE, 2009).

The babassu, an oil palm (Orbignya martiana) of great commercial and industrial value, is found in extensive natural formations in the states of Maranhão and Piauí, responsible for more than 90% of the country's production. One of the most valuable palm trees in Brazil, babassu can reach a height of 20 meters and has a set of long leaves, more than six meters long. The fruit is almond-shaped and can reach 15 cm in diameter at its widest part. The raw material used to make margarine, coconut lard, soap and cosmetics is extracted from babassu. The residue from extraction, called "babassu cake", is useful as fodder for cattle. The bud provides good quality palm heart and the fruit, while still green, is used by rubber tappers to smoke rubber (MRE, 2009).

When they ripen, their outer parts are used as food. The stem is used in rural construction and the leaves are used to make roofs for houses and baskets for the domestic industry. They can also be used to make pulp and paper. As with other types of palm, a liquid can be extracted from the cut stalk which, when fermented, produces an alcoholic drink much appreciated by the indigenous people of the region (MRE, 2009).

There are at least 2.8 million hectares of land in Maranhão, excluding current uses and protected areas, where a wide variety of oilseed crops can be planted. Also excluded from this amount are the areas defined as priorities for sugar cane, a total of 1.2 million ha (MARANHÃO, 2008). Considering an average productivity of 350 liters of vegetable oil per hectare and that half of this area (2.8 million hectares) is in a favorable logistical position, the production potential would reach 490 million liters of biodiesel per year, around 90% of the entire consumption potential of the Northeast, considering the B10 blend (10% biodiesel in 90% diesel) (MARANHÃO, 2008).

Table 5 shows the yields of the main oilseed plants in Maranhão, based on the average yields of cultivated crops obtained in 2005 (no more recent official data was found in the literature). With regard to babassu, the productivity of grains and kernels is 117 kg/ha. Considering that the grain contains 66% oil, the oil yield corresponds to approximately 77.22 kg/ha. The productivity of castor beans refers to the neighboring state of Piauí, since Maranhão has no official record of this crop in the state so far. Based on current productivity indices, the oilseed crop that stands out most in terms of yield per unit area is soybeans, with 478.8 kg/ha.

Table 5: Productivity of oilseed plants.

Culture	Productivity (kg/ha) - (grains and kernels)	Oil content (%)	Oil Yield (kg/ha)
Cotton	1.500	18,00%	270,0
Soy	2.660	18,00%	478,8
Castor beans	830	45,00%	373,5
Babassu	117	66,00%	77,22

Source: Maranhão, 2008.

4.1.1 Prospects for biodiesel production in the state of Maranhão

4.1.1.1 Biodiesel from soybean oil

Tropical soybean varieties, adapted to low latitudes, are zoned for almost the entire state, except the coast, as can be seen in Figure 9.

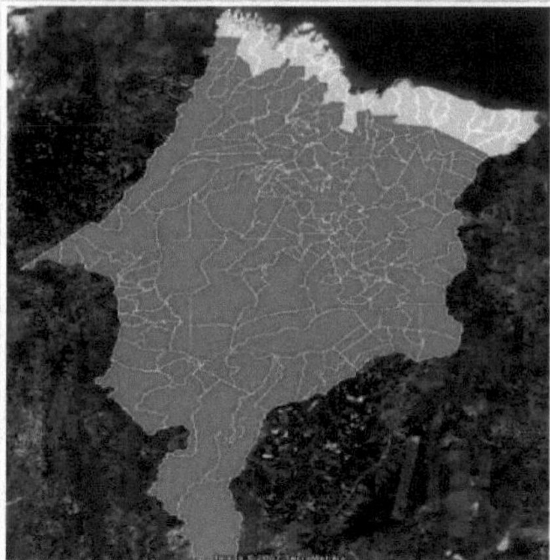

Figure 9: Availability of areas suitable for soy cultivation.
Source: Maranhão, 2008.

It is estimated that in 2010 the European Union will have the capacity to produce approximately 9.0 billion liters of biodiesel, while the projected demand is 12.0 billion liters of biodiesel to meet the targets set, according to Oil World. This would generate an import demand of 3.0 billion liters per year, which could partly be met by the state of Maranhão, given the competitive logistics of the port of Itaqui as well as its tradition in operations with the Rotterdam route, the most important in the export of minerals (MARANHÃO, 2008).

There are currently two biodiesel plant projects in the state, registered with the ANP and the Ministry of Agrarian Development, one of which is located in São Luís, in the Itaqui port area, and the other in the Porto Franco Industrial District. The total production capacity of these two projects is 141 million liters per year (MARANHÃO, 2008).

Soy is currently the oilseed with the greatest potential to be converted into biodiesel in the state. Based on production and productivity data from 2005, the potential for producing vegetable oil for a total volume of 996.9 thousand tons of grains is 179.4 million liters per year (MARANHÃO, 2008).

The installed oil extraction capacity is 90 million liters and is all concentrated in the state's only crushing plant, located in Porto Franco, next to the North-South Railway's cargo integration terminal (MARANHÃO, 2008).

Soybean production is spread across three main regions (Figure 10): Balsas Pole, located in the south of the state, with 94% of the cultivated areas; Chapadinha Pole, located in the northeast, with 4.5%; and Grajaú Pole, further towards the center of the state, with 1.5%.

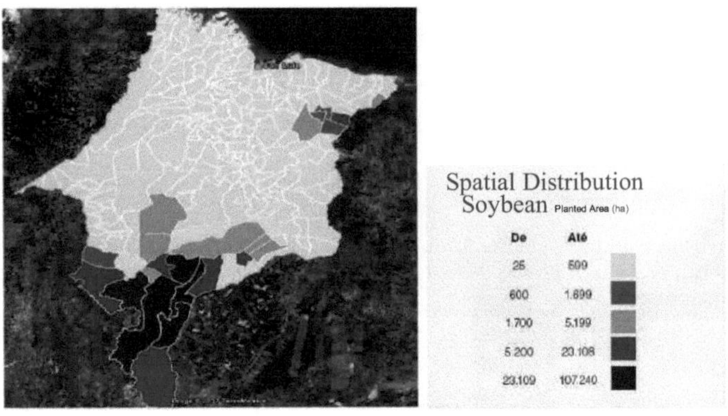

Figure 10: Distribution of soy in the state of Maranhão.

Source: Maranhão, 2008.

4.1.1.2 Cottonseed oil biodiesel

The second most important oilseed crop in the state is cotton. The area planted in 2005 was 8,385 ha, mainly in the municipalities mentioned in Table 6. Oil productivity is 270 L/ha, so total oil production was 2.26 million liters per year.

Table 6: Distribution of the area planted with cotton.

Municipalities	Planted area (ha)	Quantity produced	Average yield (kg/ha)
Tasso Fragoso	4.000	14.000	3.500

Balsas	3.984	14.342	3.599
Riachão	218	785	3600
São Felix de Balsas	120	36	300
Buriti Bravo	57	37	649
Mirador	6	6	1.000

Source: Maranhão, 2008.

The Ministry of Agriculture's agro-climatic zoning indicates that practically the entire state, with the exception of the coast, is suitable for planting cotton (MARANHÃO, 2008), as illustrated in Figure 11.

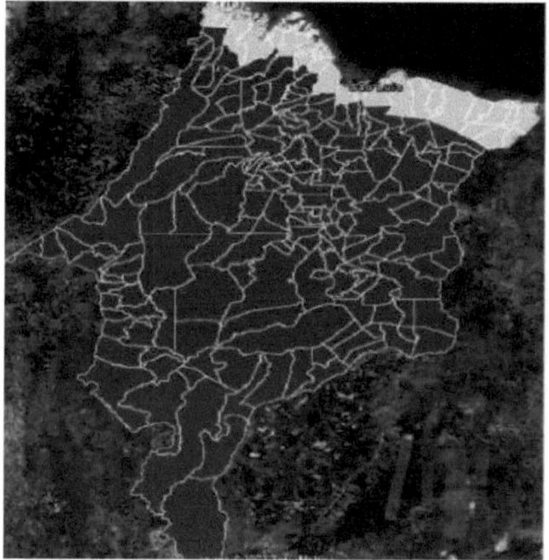

Figure 11: Distribution of areas suitable for growing cotton in the state of Maranhão.

Source: Maranhão, 2008.

4.1.1.3 Castor oil biodiesel

Although Maranhão does not have a registered area for planting this oilseed in its territory, there are at least two plant projects registered with the ANP and the Ministry of Agrarian Development, which intend to use it as a source of vegetable oil supply (MARANHÃO, 2008).

Castor bean has the characteristic of being a crop that can be exploited by family farmers, especially in the context of the settlement projects which, in Maranhão, number 875, with 97,403 families, covering a total area of 4.4 million hectares (MARANHÃO, 2008).

Biodiesel projects that combine two or more sources of vegetable oil supply, originating on the one hand from large-scale agribusiness production systems, and on the other from the crops of family farmers in settlement perimeters, are the most interesting for the state.

In anticipation of the interest that this crop is arousing in the wake of the global race for biofuels, the Ministry of Agriculture has already defined the agro-climatic zoning for the state of

Maranhão, covering 34 municipalities in the northwest and center-west of the state (MARANHÃO, 2008), as illustrated in Figure 12.

Figure 12: Availability of areas suitable for castor bean cultivation in the state of Maranhão.

Source: Maranhão, 2008.

4.1.1.4 Biodiesel from babassu oil

Three aspects have been responsible for limiting the economic importance of this activity: the manual breaking of the coconut (yield of 10 kg of kernels per person/day); the dispersion of the trees in a pattern that makes logistics difficult; and the low yield of oil per hectare compared to other oilseeds (soya: 478.8 L/ha; castor bean: 373.5 L/ha; cotton: 270 L/ha).

The expectation is that the current importance of energy biomass will provide an opportunity to revive the former economic importance that this palm once had for the state of Maranhão. The mechanization of kernel extraction is the main bottleneck to be solved, but studies indicate several work fronts with ongoing actions to create machinery that will allow for the modernization of the production process, which for decades has been based on primitive production relationships associated with extractivism and barter. The synergy provided by the economic exploitation of the co-products (endocarp charcoal, mesocarp starch) could largely compensate for the small oil yield per hectare (MARANHÃO 2008). The 4.0 million hectares of babassu palm scattered throughout the state have a potential for extracting 300 million liters of oil per year.

Currently
babassu oil production in Maranhão corresponds to around 69 million liters, 93.43% of national production, which is equivalent to 73 million liters (Table 7). However, this oil extraction corresponds to only 24.3% of the region's potential.

Table 7: Quantity and value of plant extraction products, by major region and state.

Major Regions and Federation Units	Babassu (almond)	
	Quantity (t)	Value (1,000 R$)
Brazil	110 636	115 636
North	387	391
Amazonas	12	13
Pará	30	33
Tocantins	345	345
North East	110 248	115 246
Maranhão	104 479	109 140
Piauí	5 070	5 425
Ceará	359	405
Bahia	341	275

Source: IBGE, 2009.

4.1.2 The social importance of babassu

All babassu is produced by the extremely poor people of the Amazon region, who extract the stones from the woody fruit, a laborious job due to the absolute lack of other opportunities. The kernels are sold to factories for oil extraction or they are processed into oil at home for family consumption (CLEMENT, 2005).

In most of the cultivation systems associated with babassu, the farmers don't own the land, i.e. they don't control it. Access to it is through agreements with the landowners. In general, the families of small producers are given housing, a small piece of land to cultivate and the use of extractive resources, in exchange for payment of an income in kind, usually rice (ALBIERO, 2007).

Under this system, the right to use babassu is linked to access to land. Breaking the coconut to extract the kernels is considered "women's work" (Figure 14). Considering the total time dedicated by a family to the extraction of almonds, 81% of the work is done by women (Figure 11) and children (Figure 15), while in agriculture the participation of men is greater. They also extract cooking oil, soap and a substitute for kerosene used in lamps. Charcoal is produced from the woody endocarp, which houses the kernels, and is used by families to meet their fuel needs, thus helping to protect the forests from excessive collection of wood for energy purposes (ALBIERO, 2007).

Figure 14: Babassu coconut breakers in Maranhão.

Source: MIQCB, 2009

Figure 15: Child labor in the babaçuais in Maranhão.

Source: UOL, 2009.

Zylbersztajn and colleagues (2000) reported that on average, a coconut breaker extracts around 5 kg of kernels in a day's work, although some people manage to extract up to 15 kg. There is a strong gender issue linked to the activity of breaking babassu. Because it is the only source of income generated exclusively by women in the family, the work of breaking babassu has acquired a connotation of freedom in the female imagination. The struggle of the babassu coconut breakers for improvements in the exercise of their activity gave rise to two important organizations in rural Maranhão: the Association of Rural Women Workers (AMTR) and the Interstate Movement of Babassu Coconut Breakers (MIQCB). Babassu is fully utilized by the families who survive on subsistence farming associated with the exploitation of the palm tree. The kernels that are not sold are used to produce oil and coconut milk for domestic consumption.

The mesocarp of the coconut is used for both human food and animal feed. Charcoal is produced from the endocarp and used as fuel for cooking food. The dried leaves (straw) are used to make roofs for houses. Around 5% of the kernels collected are used for domestic consumption by rural families. The rest is sold in exchange for foodstuffs (ALBIERO, 2007).

May (2000) states that as well as using the bark to produce charcoal, families use the mesocarp, which contains 60% starch, making it an excellent source of carbohydrates for feeding pigs and poultry. In times of scarcity, the mesocarp flour (cornmeal) is used as food by rural populations. Adding water to the flour produces porridge, which is considered an excellent remedy for gastrointestinal problems.

The lower the family income, the greater the relative importance of income from babassu. This suggests, according to May (2000), that any change affecting access to palm trees has serious consequences for the well-being of the population dependent on the income generated by extractive activities.

The differences in the relative importance of income from babassu depend on the economic situation of the producers. Thus, for salaried workers, the proportion of income from the sale of almonds represents 25% compared to only 11% for smallholders and squatters (MAY, 2000).

4.1.3 The economic importance of babassu

Although its exploitation is based on primary extraction, until the mid-1980s babassu played an important role in the economy of the state of Maranhão, as the basis for sustaining an industrial park for the extraction of vegetable oil, set up exclusively to process the almonds extracted from its fruit (ALBIERO, 2007).

The babassu economy reached its peak in the 60s and early 80s. During this period, 52 medium and large companies operated in Maranhão, producing oil to supply the food, hygiene and cleaning industries in the country and abroad (HERRMANN et al, 2001).

The proportion of income derived from the sale of almonds corresponds to approximately 30% of family income. This income is especially important during the off-season for annual crops, when it accounts for 42% of all the money earned. The proportion drops to 6% during the period when labor is most needed to harvest rice and there is a growing shortage of accessible babassu fruit. Although most of the extracted almonds are sold, a small proportion (5%) is destined for domestic use. The local population obtains coconut milk from the kernels, which is used in the preparation of meats, sweets and also in pure drinks or mixed with coffee (MAY, 2000).

Soybean oil is babassu oil's main competitor in the edible oil market. The formal Brazilian market for edible babassu oil is estimated at 5,500 tons/year, predominantly for the northeastern market. In addition to this, there is an informal market characterized by the self-consumption of low-income families located in the regions where the palm tree occurs. The Brazilian market for lauric oils is currently the main market for babassu oil. The hygiene, cleaning and cosmetics industries absorb 35,000 tons of crude babassu oil a year. In addition to the lauric oil market, babassu has begun to acquire importance for some companies in the steel industry, which are interested in the possibility of using carbonized coconut as charcoal to replace charcoal from native forests (ZYLBERSZTAJN et al, 2000).

The coconut shell, properly prepared, provides an efficient charcoal, an exclusive source of fuel in several regions of northeastern Brazil. The population, who know how to take advantage of the riches they have, often produce babassu charcoal at night: burned slowly in fire pits covered with leaves and earth, the babassu husk produces a vast smoke that is used as an insect repellent (Biodiesel BR, 2006).

The sale of babassu kernels represents gross revenues of US$ 18.40/ha for cattle ranchers, which in net revenues represents US$ 4.60/ha per year. Although these revenues seem low when compared to the income from cattle ranching (US$ 15.45), they represent an additional income of almost a quarter of the net income per hectare from both activities. In addition to being an important source of income for the local people who extract it, babassu is also an important source of income for the regional economy as a whole, due to the industrialization of babassu oil, which in the 1980s reached a production of 80,000 tons/year, generating a final market value of 40 million dollars (MAY, 2000).

When babassu oil is transformed into biodiesel (Figure 16), it can be a viable economic alternative, since the biofuel can be produced at the place of use, an interesting configuration especially for the isolated regions of our country (LOPES, 1983).

Wunder (1998) stated that the extractive subsector of agriculture has received a lot of international attention because of the potential it has been given for the sustainable use of tropical forests and other natural ecosystems, for example, such as the harvesting of non-timber products in extractive reserves. In this context, babassu receives special attention, as it is one of the raw materials that fits into this context.

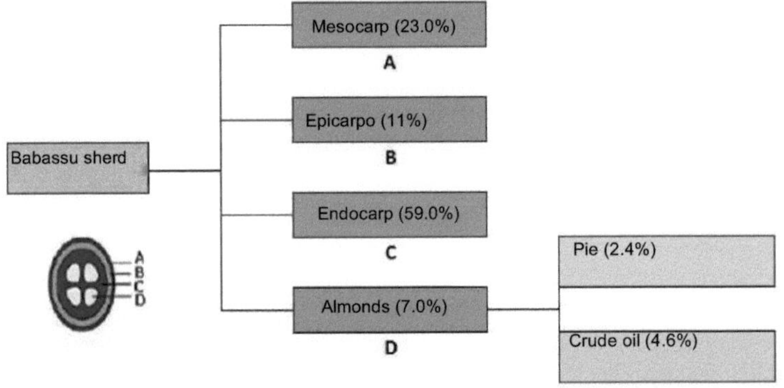

Figure 16: Products for industrializing babassu coconut.
Source: Pensa 2000; PARENTE, 2003 ; DESER, 2005, apud SANTOS 2008.

4.1.4 Cooperative organization in the Médio Mearim region - ASSEMA

Based on the research carried out, a settlement association called "Associação em Áreas de Assentamento no Estado do Maranhão" - ASSEMA - was identified in the Médio Mearim region.

ASSEMA is an organization led by rural workers and women babassu coconut breakers, which promotes family production, using and preserving babassu groves to improve the quality of life in the countryside. ASSEMA is a regional non-profit organization operating in the so-called Médio Mearim region, in the state of Maranhão, located in the Mid-North of Brazil (ASSEMA, 2009).

ASSEMA was founded in 1989 on the initiative of union leaders from the municipalities of Esperantinópolis, Lima Campos, São Luiz Gonzaga do Maranhão and Lago do Junco, who wanted to invest in improving working and living conditions and defending the babassu plantations.

For 20 years, ASSEMA has been providing technical, legal, economic and political advice to rural workers' families, especially babassu coconut breakers, in their efforts to improve the economy of family farming, encouraging the organization of cooperative and associative systems for the production of organic food (ASSEMA, 2009).

ASSEMA has contributed to:

- Strengthening the right to citizenship for families of rural workers and babassu coconut breakers.
- The creation of sustainable and solidarity-based economic initiatives.
- Promoting equity in gender, generational and ethnic relations.
- Raising awareness of the importance of organic ecological agriculture.
- Combating the rural exodus with alternatives that help people stay on the land they have conquered.
- The implementation of public policies that prioritize the preservation of babassu groves and municipal laws that guarantee coconut breakers free access to babassu groves.

ASSEMA's vision is to promote sustainable development in order to enable the autonomy of settled families. ASSEMA's mission is the collective construction, by the rural workers and babassu coconut breakers of Médio Mearim, of sustainable actions for the use of natural resources in the search for quality of life in the countryside, based on family production, fair gender relations and respect for ethnicities and cultural diversity (ASSEMA, 2009).

Its membership includes 28 organizations, representing around 2,500 families living in 43 communities. Among the people who founded ASSEMA, there are also those who founded COPPALJ - Cooperativa dos Pequenos Produtores Agroextrativistas de Lago do Junco and other associations and cooperatives in the Médio Mearim, such as the coconut breakers who founded AMTR - Associação de Mulheres Trabalhadoras Rurais do Lago do Junco e Lago dos Rodrigues and COOPAESP - Cooperativa dos Pequenos Agroextrativistas de Esperantinópolis (ASSEMA, 2009).

As ASSEMA possesses the characteristics sought in this research, in the sense of enabling financing conditions for the installation of a biodiesel production unit, and therefore improving the socio-economic condition of the region, this association was considered a model cooperative in this study of the implementation of the biodiesel plant installation project.

4.1.5 Division of the state of Maranhão into regions

The regional diversities of the Brazilian territory, particularly Maranhão, are the result of a complex and dynamic matrix of interacting environmental, social, economic, cultural and political factors. In order for public policies to be effective, it is essential for society to participate in the process of drawing them up and implementing them, taking into account this set of factors that make up the diversity of Maranhão's territory (SEPLAN et al, 2008).

The government of the state of Maranhão, with Complementary Law No. 108 of 21 November 2007, created the new regionalization with the division of the state into 32 planning regions, as mentioned in Figure 17. The deconcentration of the administrative structure and the implementation of decentralized planning is aimed at the sustainable development of the regions, through the strengthening of municipalities and partnerships with organized civil society (SEPLAN et al, 2008).

With this initiative, public policies should be implemented more effectively and concretely, in a new state-territory relationship, aimed at enhancing the potential of the regions, reducing inequalities and improving the quality of life of the population of Maranhão (SEPLAN et al, 2008).

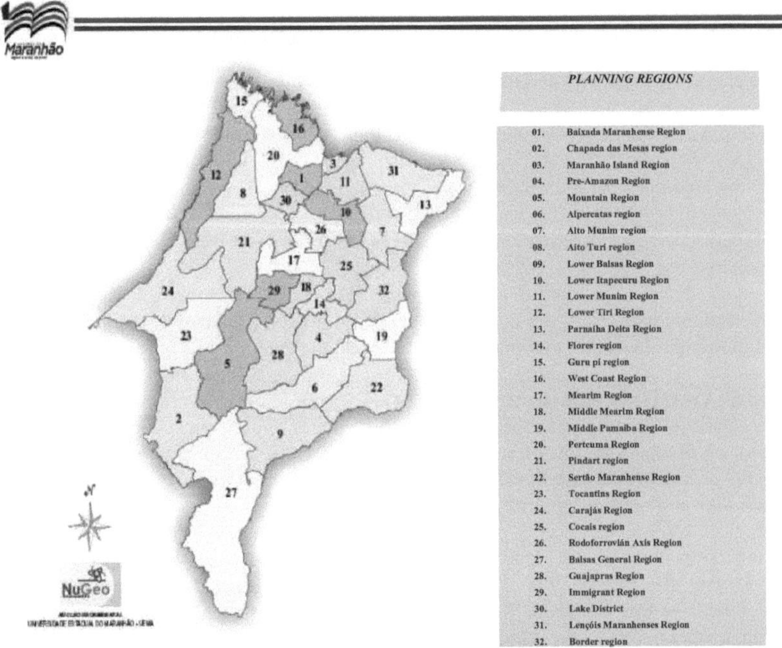

Figure 17 : Planning regions of the state of Maranhão.

Source: SEPLAN, 2008.

Within the summary shown in Table 8, it is clear that the state of Maranhão has several regions with lower socio-economic indicators than other regions in Brazil. Designing and implementing strategies to increase these indicators is one of the major challenges facing government leaders.

Table 8: Summary of Maranhão's socio-economic indicators.

Caption
Higher value
Intermediate Value
Lower value

Region	Population Resident - 2007	GDP 2006 (in R$)	Per capita GDP 2006 (in R$)	% of the number of formal jobs over the total population aged 18 to 64 - 2007	% of people benefiting from the bolsa família program
Brazil	183.987.291	2.369.796.546	12.688,28	31,57	18,91
North East	51.534.406	311.174.975	6.029,47	19,83	37,05
Maranhão	6.118.995	28.621.445	4.628,20	14,15	45,27
Baixada Maranhense Region	108.761	212.445	1.924,90	3,58	58,83
Chapadas das Mesas region	111.425	545.415	4.920,17	13,27	42,29
Maranhão Island Region	1.211.270	11.781.159	9.380,81	32,08	24,91
Pre-Amazon Region	154.047	433.605	2.733,25	5,76	52,11
Mountain Region	126.992	356.260	2.770,21	8,77	49,03
Alpercatas region	110.392	293.326	2.594,93	4,45	57,01
Alto Munim region	169.621	484.596	3.079,75	7,08	61,50
Alto Turi Region	109.084	331.206	2.950,64	5,92	56,39
Lower Balsas Region	48.832	310.949	6.427,74	13,61	52,76
Lower Itapecuru Region	172.425	432.453	2.785,27	7,10	51,27
Lower Munim Region	128.825	303.374	2.622,73	6,83	54,19
Lower Turi Region	87.056	314.566	3.229,97	5,19	52,63
Parnaiba Delta Region	166.725	380.967	2.281,73	4,84	61,84
Flores region	95.318	258.675	2.707,91	6,35	56,98
Gurupi region	62.397	151.496	2.510,38	6,32	47,64
West Coast Region	125.526	257.844	2.040,40	4,51	54,94
Mearim Region	217.011	735.831	3.425,28	7,68	55,91
Middle Mearim Region	128.210	466.328	3.309,90	8,96	54,78
Middle Parnaiba Region	207.523	587.573	2.842,57	11,08	43,76
Pericumã Region	252.378	566.292	2.321,26	4,99	54,32
Pindaré Region	347.786	1.250.003	3.370,11	7,66	52,91
Sertão Maranhense Region	128.172	274.704	2.213,42	5,85	53,05
Tocantins Region	360.585	1.767.021	4.793,74	17,38	34,68
Carajás Region	248.063	1.973.666	7.567,74	16,23	38,16
Cocais region	238.314	692.465	2.780,39	7,78	57,57
Road and rail hub region	158.921	377.899	2.392,71	5,87	54,20
Balsas General Region	133.224	955.797	7.405,54	19,13	37,94
Guajajaras region	101.638	304.678	3.103,93	5,96	52,04
Immigrant Region	91.996	291.207	3.099,66	5,53	61,04
Lagoon Region	126.271	295.791	2.452,68	5,50	57,40
Lençóis Maranhenses Region	154.358	309.301	2.091,37	10,93	60,89
Timbiras Region	235.849	924.553	3.945,18	12,15	51,71

Source: SEPLAN, 2008.

Among the different planning regions of Maranhão, the Middle Mearim and Cocais regions were chosen for this research.

The Médio Mearim region was chosen for this study due to the existence of ASSEMA in this region, which in turn has cooperatives, an essential factor for collective financing in the PRONAF line with a limit of up to R$ 10,000,000.00. Currently, around 500 people are part of the cooperative, developing and selling handicraft products and working with oil extraction on a small scale, with annual production of around 180,000 L .

Figure 18 shows the map of the Médio Mearim region, which is made up of 9 municipalities. The population of the Médio Mearim region is 128,210, of which 44,528 - 34.73% - are located in rural areas and 83,682 - 65.27% - in urban areas. The demographic density is 41 (inhab/km²) and the GDP in 2005 was R$417.6 million (Table 9).

Figure 18: Middle Mearim region.

Table 9: Geographical, economic and social characterization of the Médio Mearim region.

Region / Municipality	Area (km²)	Population 2007			Population density (inhab / km²)	GDP 2005* R$ million
		Urban	Rural	Total		
MIDDLE MEARIM	3.114,2	83.682	44.528	128.210	41,2	417,6
Bernardo do Mearim	261,4	2.028	3.944	5.972	22,8	19,2
Esperatinópolis	480,9	9.443	9.126	18.569	38,6	43,3
Igarapé Grande	374,3	6.006	4.670	10.676	28,5	24,6
Lima Campos	321,9	6.629	4.736	11.365	35,3	23,8
Quarries	288,5	32.011	5.973	37.984	131,7	166,7
Potion of Stones	655,2	7.984	7.869	15.853	24,2	66,0
São Raimundo do Doca Bezerra	281,2	1.628	2.874	4.502	16,0	13,1
São Roberto	227,5	2.432	2.557	4.989	21,9	9,0
Trizideia do Vale	223,3	15.521	2.779	18.300	82,0	51,9

Source: IMESC (a), 2009.

Analyzing the share of economic activities in the Médio Mearim region, it can be seen that forestry and logging, to which babassu extraction is added, accounts for 58% of the added value. This activity therefore makes a significant contribution to the region's economy (Table 10).

Table 10: Share of economic activities in the added value of agriculture and livestock in the Middle Mearim region.

Region and Municipalities	Value Added (in thousand R$)				
	Total	Agriculture	Livestock	Forestry and logging and related services	Fishing
MIDDLE MEARIM	136.344	28.008	27.417	78.694	2.225
Bernardo do Mearim	12.750	1.117	4.010	7.624	
Esperantinópolis	12.099	2.988	2.700	6.076	336
Igarapé Grande	8.892	1.529	2.168	5.196	
Lima Campos	6.816	3.656	1.069	2.091	
Quarries	36.343	7.380	4.510	23.418	1.034
Potion of Stones	37.327	4.404	7.842	24.980	101
São Raimundo do Doca Bezerra	5.743	1.865	1.306	2.404	168
São Roberto	3.360	1.264	1.020	1.346	
Trizidéia do Vale	12.743	3.804	2.793	5.560	586

Source: IMESC (a), 2009.

The Médio Mearim region has a large concentration of babassu kernels, approximately 15,000,000 kg in 2006. Taking into account that approximately 66% of the kernel is made up of oil, the production capacity for this concentration of kernels is 9,900,000 L of oil (Figure 19).

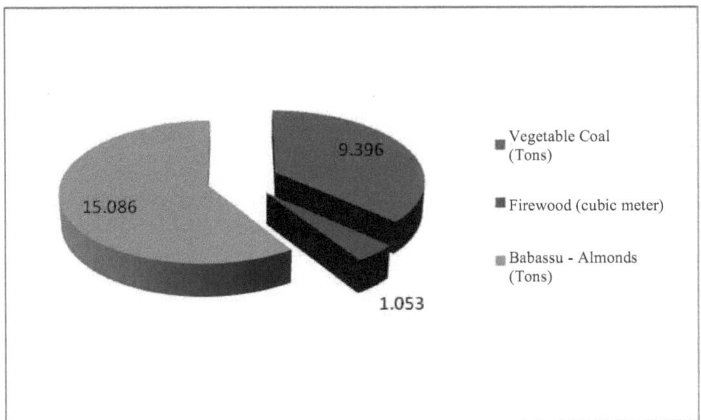

Figure 19: Quantity of the main plant extraction products in tons and cubic meters in the Médio Mearim region.

Source: IMESC (a), 2009.

When it comes to temporary crops, the Médio Mearim region concentrates its agricultural activities on rice and corn, which are essential for food. It can be seen that soy, one of the raw materials that could be used to make biodiesel, is not represented in the region (Table 11).

Permanent farming is concentrated on banana cultivation, which is the region's greatest potential (Table 12).

The population living in the Médio Mearim region is young (Figure 20) and could be used as labor in the manufacture of biodiesel and by-products. Another interesting point is the investment the government could make to qualify these people through technical courses in related areas, such as chemistry, physics, administration, etc. Table 11: Main agricultural products (temporary crops), by quantity produced, in the Médio Mearim region.

Region and municipalities	Rice	Sugarcane	Beans	Cassava	Corn	Pineapple	Watermelon	Tomato
	Tons	Tons	Tons	Tons	Tons	A thousand fruits	Tons	Tons
MIDDLE MEARIM	17.413	1.255	897	1.571	6.511	108	311	62
Bernardo do Mearim	1.316		100	220	380			
Esperantinópolis	2.260	550	118	400	600		3 0	
Igarapé Grande	1.535		140	165	520			
Lima Campos	942	175	24	92	183		2 4	

Quarries	810	105	40	67	290		104	
Potion of Stones	5.890		331	168	2.980	54	3 5	62
São Raimundo do Doca Bezerra	2.280		68	200	780	54	3 0	
São Roberto	1.660		41	164	548		2 0	
Trizidéia do Vale	720	425	35	85	230		6 8	

Source: IMESC (a), 2009.

Table 12: Main agricultural products (permanent crops), by quantity produced, in the Médio Mearim region.

Region and municipalities	Coconut from Bahia	Mango	Cashew nuts	Banana	Orange	Lemon
	A thousand fruits	Tons	Tons	Tons	Tons	Tons
MIDDLE MEARIM	110	54		16.370	199	125
Bernardo do Mearim	13		15	220	12	
Esperantinópolis	12	54	3	1.220	36	
Igarapé Grande	21		5	550	32	
Lima Campos				3.400	12	
Quarries	13		5	6.300	32	125
Potion of Stones	30		2	580	28	
São Raimundo do Doca Bezerra	8			450	12	
São Roberto	5			320	13	
Trizidéia do Vale	8			3.330	22	

Source: IMESC (a), 2009.

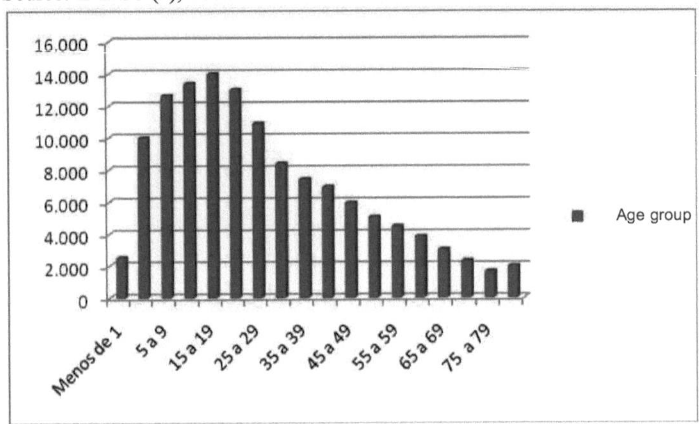

Figure 20: Resident population by age group in the Médio Mearim region.
Source: IMESC (a), 2009.

The average salary of the population in the Médio Mearim region is between 0.5 and 2 minimum wages (Table 13). There is a pressing need to increase the income of people in this region.

Table 13: Workers by income bracket, in minimum wage, in the Médio Mearim region.

Region and municipalities	0- 0,50	0,51 -1	1,01 - 2	2,01 - 3	3,01 - 5	5,01 - 10	10,01 - 15	15,01 - 20	IGNORED	MORE THAN 20
MIDDLE MEARIM	68	1.869	3.270	489	157	86	13	8	4	50
Bernardo do Mearim		41	124	19		1	1			
Esperantinópolis	1	343	127	4	1	4		2		1
Igarapé Grande		57	253	7	2	3	1	1		
Lima Campos		223	193	7	8	1	1		1	
Quarries		432	1.831	216	106	58	8	4	3	9
Potion of Stones	67	159	388	234	36	13	2	1		40
São Raimundo do Doca Bezerra		109	170							
São Roberto		112	1							
Trizidéia do Vale		393	183	2	4	6				

Source: IMESC (a), 2009.

As mentioned in Table 9, the population of the Médio Mearim region is 128,210. It is estimated that 24,485 families are poor and 19,493 are beneficiaries of the Bolsa Família program. If an average of 4 people per family is taken into account, the number of poor people is 97,940 - 76.39% of the population, of which 77,972 are beneficiaries of the bolsa família program - 60.82% (Table 14).

Region and municipalities	Estimate: Poor Families - Cadastro Único Profile (based on IBGE 2004 data)	Number of Families Benefiting from Bolsa Família - September 2008
MIDDLE MEARIM	24.485	19.493
Bernardo do Mearim	875	784
Esperantinópolis	3.750	2.995
Igarapé Grande	1.928	1.631
Lima Campos	2.243	1.820
Quarries	6.284	4.889
Potion of Stones	4.490	3.440
São Raimundo do Doca Bezerra	1.147	913
São Roberto	900	719
Trizidéia do Vale	2.868	2.302

Source: IMESC (a) 2009.

Given the panorama presented, the main demands of the Médio Mearim region are (SEPLAN, 2008):

- Expanding, restoring and maintaining the road network.
- Encouraging agribusiness.
- Developing the babassu productive arrangement.
- Encouraging tourism.
- Create a handicraft marketing center.

The installation of the biodiesel production plant in this region could indirectly contribute to at least two of the items mentioned above: agribusiness and the babassu production arrangement.

4.1.5.2 Socio-economic analysis of the Cocais region

Figure 21 shows the map of the Cocais region, which is made up of 5 municipalities. The population of the Cocais region is 238,314, of which 83,041 - 34.85% - are located in rural areas and 155,273 - 65.15% - in urban areas. The population density is 25 (inhabitants/km²) and the GDP in 2005 was R$526.4 million (Table 15).

Comparing the Cocais region with the Médio Mearim, the Cocais region has 110,000 more people - 53.8%, a lower population density and a GDP around R$108.8 million higher.

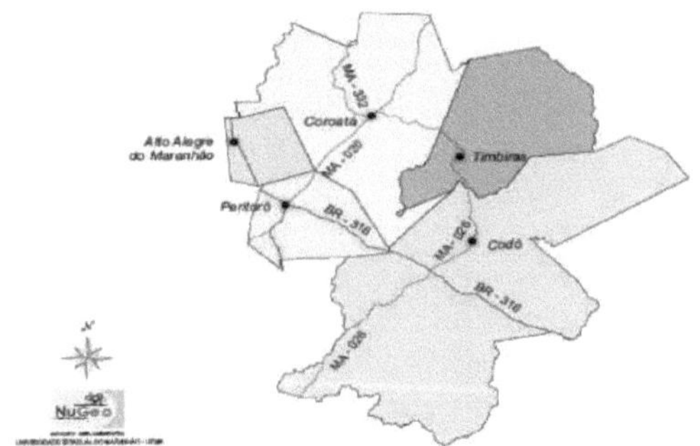

Figure 21: Cocais region

Source: IMESC (b), 2009. Table 15: Geographical, economic and social characterization of Cocais.

Region / Municipality	Area (km²)	Population 2007			Population density (inhab / km²)	GDP 2005* R$ million
		Urban	Rural	Total		
COCAIS	9.283,4	155.273	83.041	238.314	25,7	526,4
Alto Alegre do Maranhão	420,9	14.941	7.061	20.002	52,3	40,2
Codó	4.364,5	76.209	34.365	110.574	25,3	285,2
Coroatá	2.263,8	40.850	19.739	60.589	26,8	125,2
Peritoró	747,7	7.442	11.575	19.017	25,4	32,0
Timbiras	1.486,5	15.831	10.301	26.132	17,6	43,9

Source: IMESC (b), 2009.

In relation to the Médio Mearim region, the participation of economic activities is more balanced, but forestry and logging contributes approximately 40% of the added value of agriculture and livestock (Table 16).

Table 16: Participation of economic activities in the added value of agriculture and livestock in the Cocais region.

Region and Municipalities	Value Added (in thousand R$)				
	Total	Agriculture	Livestock	Forestry and logging and related services	Fishing
COCAIS	90.162	30.415	21.418	36.430	1.900
Alto Alegre do Maranhão	13.207	3.917	3.379	5.737	174
Codó	40.731	16.992	9.131	13.588	1.020
Coroatá	17.838	4.905	3.852	8.619	461
Peritoró	7.810	1.701	2.562	3.455	92
Timbiras	10.576	2.900	2.494	5.031	152

Source: IMESC (a), 2009.

In 2006, the extraction of babassu from the Cocais region corresponded to 10,450,000 kg of kernels. Taking into account that approximately 66% of the kernel is made up of oil, the annual production capacity was 6,897,000 L of oil (Figure 22).

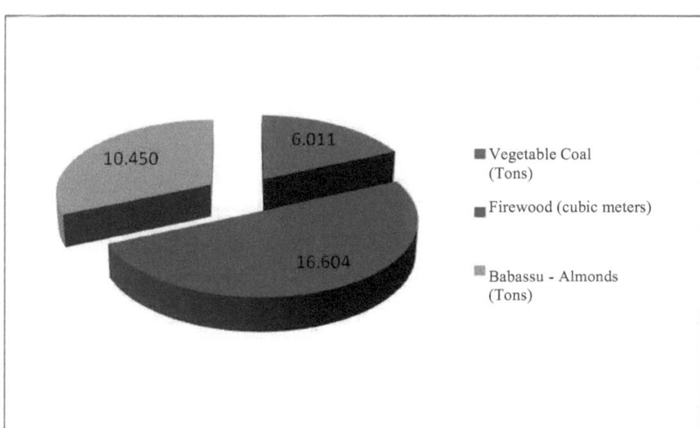

Figure 22: Quantity of the main plant extraction products in tons and cubic meters in the Cocais region.

Source: IMESC (b) 2009.

When it comes to temporary crops, the region concentrates its agricultural activities on growing cassava, rice and corn, which are essential for food, and sugar cane, which is vitally important for alcohol plants. The municipality of Codó plays an important role in the agricultural products of temporary farming in the Cocais region (Table 17).

It can be seen that soy, one of the raw materials that could be used to make biodiesel, is not represented in the region, similar to the Médio Mearim region.

Table 17: Main agricultural products (temporary crops), by quantity produced, in the Cocais region.

Region and municipalities	Rice	Sugarcane	Beans	Cassava	Corn	Watermelon
	Tons	Tons	Tons	Tons	Tons	Tons
COCAIS	29.433	13.658	786	45.600	11.794	1.041
Alto Alegre do Maranhão	2.349		277	2.500	1.308	
Codó	11.664	12.960	201	36.000	4.662	828
Coroatá	7.798	210	165	3.900	3.042	153
Peritoró	2.520	160	61	1.400	930	32
Timbiras	5.102	328	82	1.800	1.852	28

Source: IMESC (a), 2009.

Bananas and oranges are the main permanent crops in the Cocais region, while in the Médio Mearim region bananas are much more widely produced (Table 18).

Table 18: Main agricultural products (permanent crops), by quantity produced, in the Cocais region.

Region and municipalities	Coconut from Bahia	Mango	Cashew nuts	Banana	Orange	Lemon
	A thousand fruits	Tons	Tons	Tons	Tons	Tons
COCAIS	57	114	38	716	336	4
Alto Alegre do Maranhão				57	18	
Codó	6	81	35	382	225	
Coroatá	9		2	72	51	4
Peritoró		25	1	91	28	

Timbiras	42	8		114	14	

Source: IMESC (b) , 2009.

Similar to the Medio Mearim region (Figure 20), the population living in the Cocais region is young and could be used as labor in the manufacture of biodiesel and its by-products (Figure 23). The government could and should invest in training this workforce.

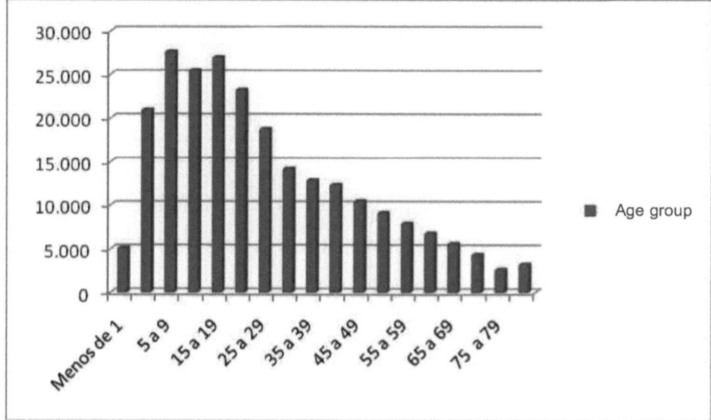

Figure 23: Resident population by age group in the Cocais region.

Source: IMESC (b), 2009.

The average salary of the population of the Cocais region was between 0.5 and 2 minimum wages in 2006, showing that there is a need to increase the income of people in the region (Table 19).

Table 19: Workers by income bracket, in minimum wage, in the Cocais region.

Region and municipalities	0- 0,50	0,51 -1	1,01 - 2	2,01 - 3	3,01 - 5	5,01 - 10	10,01 - 15	15,01 - 20	IGNORED	MORE THAN 20
	SM	SM	SM	SM	SM	SM	SM	SM		SM
COCAIS	42	3.088	5.916	465	275	104	33	13	10	14
Alto Alegre do Maranhão	29	37	667	5	1					1
Codó	9	1.402	3.484	273	200	82	22	9	9	9
Coroatá	2	1.270	1.111	128	42	18	9	2	1	4
Peritoró	1	98	444	6	8		2	1		
Timbiras	1	281	210	53	24	4		1		

Source: IMESC (b), 2009.

As mentioned in Table 16, the population of the Cocais region is 238,314. It is estimated that 45,560 families are poor and 36,729 are beneficiaries of the Bolsa Família program. If an average of 4 people per family is taken into account, the number of poor people is 182,240 - 76.47% of the population, of which 146,916 are beneficiaries of the Bolsa Família program - 61.65% (Table 20).

Compared to the Médio Mearim region, the Cocais region has 0.08% more poor families and 0.83% more families benefiting from the Bolsa Família program.

Table 20: Families benefiting from and registered with the Bolsa Família program in the Cocais region.

Region and municipalities	Estimate: Poor Families - Cadastro Único Profile (based on IBGE 2004 data)	Number of Families Benefiting from Bolsa Família - September 2008
COCAIS	45.560	36.729
Alto Alegre do Maranhão	3835	3.151
Codó	21.040	16.183
Coroatá	11.662	10.068
Peritoró	3.734	3.081
Timbiras	5.289	4.246

Source: IMESC (b), 2009.

Given the panorama presented, the main demands of the Cocais region are (SEPLAN, 2008):

- Encourage agro-industry (babassu and ceramics).
- Maintaining roads and improving railroads.
- Recovering the Itapecuru basin.
- Strengthening the craft production chain.
- Building health centers in quilombola territories.
- Build a medium and highly complex regional hospital.
- Install a solid waste recycling plant.

4.1.5.3 Identifying the region for setting up a biodiesel production unit in Maranhão

The region chosen for the installation of the biodiesel plant was the Médio Mearim region, where ASSEMA and the cooperatives are located, an essential factor for financing from the PRONAF line, as well as the strong presence of babassu coconut. In order to set up the biodiesel production unit, the possibility of acquiring the raw materials for biodiesel production from family farming was considered, since ASSEMA is seen as a model for this study. The use of raw materials generated by family farming aims to contribute to social inclusion and regional development by generating employment and income for family farmers who fall within the PRONAF criteria. By using this method, the biodiesel producer will be entitled to the social fuel label.

The relevance of the Cocais region for this study is due to its proximity to the Médio Mearim region and the representativeness of babassu coconut extraction. Therefore, if there is a need to expand biodiesel production, or even contingencies in the acquisition of raw materials, the Cocais region will have a strong contribution to make in these areas.

4.1.5.4 Potential of the biodiesel produced by the plant in relation to the Brazilian market

The ANP has been holding biodiesel auctions since 2005. At the auctions, refineries and distributors buy biodiesel to mix with petroleum-based diesel. The initial aim of the auctions was to generate a market and thus stimulate the production of biodiesel in sufficient quantities for refineries and distributors to be able to make up the mixture determined by law.

Auctions continue to be held to ensure that all diesel oil sold in the country contains the percentage of biodiesel determined by law. The production and use of biodiesel in Brazil promotes the development of a sustainable energy source from an environmental, economic and social point of view and also brings the prospect of reducing diesel oil imports.

The 16th biodiesel auction took place on November 17, 2009 (Table 21), in Rio de Janeiro, when 575 million liters were traded, guaranteeing the B5 blend (5% biodiesel in diesel), which was sold on January 1, 2010 (ANP, 2009).

Table 21: 16th biodiesel auction.

Description	Auction 81/09
Offering units	40
Classified units	40
Volume offered in (m^3)	725.179
Volume classified/awarded (m^3)	575.000
Maximum reference price (m^3)	2.350,00
Lowest price (R$ / m^3)	2.270,00
Average price (R$ m^3)	2.326,67
Maximum discount	- 3,40%
Average discount	- 0,99%
Amount traded	1.337.837

Source: ANP, 2009.

The volume auctioned represents the entire demand for biodiesel for this blend in the first quarter of 2010. Compared to the previous auction (for the B4 blend), it represents a 25% increase in production, which benefits the entire biofuel production chain (ANP, 2009). The states that won the largest volumes of biodiesel were Mato Grosso and Rio Grande do Sul (Table 22).

Table 22: Volume auctioned by state in (m^3).

UF	Producer	Volume (m)[3]
BA	B. Ecodiesel - Iraquara	13314
	Petrobras - Candeias	16500
BA-Total		29.814
EC	B. Ecodiesel - Cratéus	100
	Petrobras - Quixada	15600
CE-Total		15.700
GO	Bi natural	15000
	Caramuru	34 330
	Bulk - Anápolis	44 100
GO-Total		99.430
MA	B. Ecodiesel - São Luis	19000
MA-Total		19.000
MG	B-100	2 160
	Petrobras - Montes Claros	16500
MG-Total		18.660
MS	Biocar	2 160
MS-Total		2.160
MT	ADM	68760
	Agrosoia	4400
	Araguassú	900
	Barrálcool	8910
	Bio Oil	720
	Biocamp	10350
	Biopar (Parecis)	1 680
	CLV	7.070
	Cooperbio	19000
	Fiagril	25 500
	SSIL	360
	Transp Caibiense	1 080
MT-Total		148.730
PA	Agropalma	1 300
PA-Total		1.300

UF	Producer	Volume (m)[3]
PA	Agropalma	1.300
PA-Total		1.300
PI	B Ecodiesel - Floriano	100
PI-Total		100
PR	Biopar - (Rolândia)	6.930
PR-Total		6.930
RJ	Cesbra	4.300
RJ - Total		4.300
RO	Amazombio	3.200
RO - Total		3.200
RS	B Ecodiesel - R do Sul	14.850
	BS Bios	30.000
	Granol - Cacheoira do Sul	
	Oleopan	44.300
RS - Total		125.050
SP	Biocapital	30.500
	Bioverde	17.640
	Bracol	23.230
	Fertibom	8.520
	Innovati	2.160
	SP-Bio	4.996
SP-TOTAL		87.046
TO	B Ecodiesel - National P	18.000
	Biotins	1.580
TO-TOTAL		19.580
Grand Total		575.000

Source: ANP, 2009.

Figure 24 shows biodiesel sales by region. It can be seen that the Northeast region, the subject of this study, is not very representative.

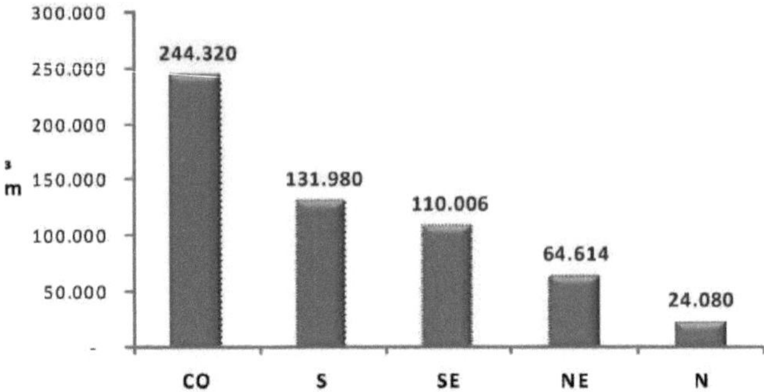

Figure 24: Volume auctioned by region in (m³).
Source: ANP, 2009.

Table 23 shows the projected demand for biodiesel, which is made up of B2 from January 1, 2008, B3 from July 1, 2008 and B5 from 2010. The scenario is positive, due to the increase in demand for biodiesel (MME, 2009).

It should be noted that B5 was brought forward to 2010, and as of January 1, 2010, it became compulsory to mix 5% biodiesel in all diesel oil consumed in Brazil, except marine diesel oil. According to current market data, the new blend should generate foreign exchange savings of around US$ 1.4 billion/year due to the reduction in diesel oil imports (MME, 2010).

The demand for biodiesel in the Northeast region for 2010 is 361,000,000 L of biodiesel. The plant planned in this dissertation will produce 3,600,000 L of biodiesel per year, corresponding to only 1% of the demand. To meet all of the region's demand, approximately 100 biodiesel plants would be needed, just like the one envisioned in this research.

Table 23: Biodiesel demand by region, in (thousand m³) 2008 to 2017.

Year	North	North East	Midwest	South East	South	Brazil
2008	116	162	119	515	219	1.130
2009	137	206	149	648	276	1417
2010	240	361	262	1.130	484	2.477
2011	252	379	276	1.186	510	2602
2012	237	399	291	1.246	536	2.709
2013	248	419	307	1.310	565	2.849
2014	261	442	323	1.377	595	2.999
2015	274	466	341	1.448	626	3.156
2016	288	492	359	1.524	660	3.322
2017	302	518	378	1.603	695	3.495
Increase (thousand m³)						
2008 - 2017	281	456	354	1.493	653	3.236
Increase (% in year)						
2008 - 2012	19,6	25,3	25,1	24,7	25,1	24,4
2012 - 2017	5,0	5,4	5,4	5,2	5,3	5,2
2008 - 2017	11,2	13,8	13,7	13,4	13,7	13,4

Source: EPE, 2009.

4.2 ECONOMIC ANALYSIS OF THE BIODIESEL PLANT IN MARANHÃO

Economic analyses of technologies for different scales in order to study economic viability have often been based on production volume. This process is almost always done without taking into account the changes in product quality during the change in scale. The use of linear relationships in most cases leads to technically and financially incorrect decisions. This is due to the relationship between product quality and sales price. This situation requires knowledge of the true scale relationships between plants of different sizes in order to understand the possible distortions between the quality characteristics of the product or process at different scales and their influence on economic and financial parameters.

The regions of Maranhão that were assessed in this study with regard to economic issues were the Médio Mearim and Cocais regions, as described in items 4.1.5.1 and 4.1.5.2. Both the Médio Mearim and Cocais regions have great babassu potential. Using information from the IBGE in 2006, the two regions together processed 25,536,000 tons of babassu kernels, which is equivalent to approximately 16 million liters of oil per year.

For this study, a plant with a production capacity equivalent to 300,000 liters per month was considered, which corresponds to an annual production of 3,600,000 liters. As the two regions have a production capacity of approximately 16 million liters of babassu oil, there will be no shortage of raw materials for the planned production.

4.2.1 Economic viability of the biodiesel production process based on the commercial proposal

Two commercial proposals were analyzed that have distinct peculiarities, since they were drawn up using different technologies, but due to the confidentiality requested by them, the proposals could not be presented in this research, so wc will use the representation of manufacturer A and manufacturer B.

Similarly, the amounts budgeted were different depending on the technology used by the companies.

Manufacturer A's commercial proposal was used in the research, as it is more accessible for assessing the economic viability of the project to produce 300,000 liters of biodiesel per month.

A project's investment plan is a detailed description of the capital needs required for the project to materialize. A project's investments are intended for two purposes: one refers to fixed investments and the other to what is known as working capital. Fixed investments are made during the installation period of the project and are used during the useful life of the corresponding assets. These investments include assets that are subject to depreciation, such as machinery, vehicles, buildings, land, furniture, etc. They also include non-weighted investments, which are also subject to depreciation, such as project studies, organizational expenses (technical and legal advice, travel expenses, among other expenses).

Working capital, in turn, corresponds to the investments the company must make to meet its operational needs during a year of the plant working one shift, totaling 12 hours and operating 25 days a month, to produce biodiesel, its most important product. At a later date, the plant can operate in two shifts totaling 24 hours.

Working capital was divided into two components: one fixed, mainly related to plant maintenance, and the other variable, primarily involving the input used in the process.

The fixed capital consists of a metal structure, storage tanks, main reactor, catalysis reactor, decanter tank, distiller, hydraulic pumps, pipes, electrical system, automation system, measuring equipment, quality control equipment, bag filter, filter press and centrifuge. In manufacturer A's commercial proposal, this equipment was valued at R$3,553,000.00. Babassu coconut depulping machines will also be purchased for processing babassu coconuts, with a unit price of around R$12,000.00. As the capacity of the dehusking machine is approximately 400 kg of kernels every 12 hours and the biodiesel plant will use 13,200 liters of babassu per day, 50 machines will be needed for this capacity, for a total price of R$ 600,000.00. It is also necessary to purchase an oil extraction press, which will cost approximately R$ 450,000.00 per unit, with a capacity of up to 2 tons of kernels per hour - approximately 1,320 liters of babassu oil. Therefore, the total value of the fixed capital investment is R$4,603,000.00. Assuming that the average useful life of this equipment is 10 years, its annual depreciation is R$ 460,300.00.

4.2.2 Fixed costs

Fixed costs are those that do not vary with the company's level of activity. Fixed costs are therefore independent of the quantity produced. It is clear that they fluctuate periodically, either as a result of adjustments to the personnel structure or due to better administrative rationalization. Fixed costs can be divided into three categories: personnel, administrative and financial.

- Personnel: Wages and salaries, social charges, benefits (health insurance, meal and transport vouchers, life insurance, etc.).
- Administrative: Rent, taxes, accountant, advisors (financial, lawyer, environmental, etc.), electricity, telephone, maintenance of vehicles and equipment, insurance of this equipment and vehicles, advertising, travel expenses, and other expenses (including interest and bank fees).
- Financial: Interest associated with industrial facilities.

The taxes to consider are as follows: Tax on the Circulation of Goods and Services (ICMS); Social Integration Program (PIS); Social Financing Contribution (COFINS); and Income Tax on Real Profit.

The operating costs, which correspond to almost all the fixed operating costs (maintenance work, overheads, insurance, among others), were based (PAGLIARDI, 2004; MAX, 1991) on a percentage of the initial investment (Table 24).

Table 24. Annual fixed operating costs (10-hour period/25 days of the month).

Fixed costs - monthly and annual figures			
Description	% of Initial Investment	Monthly amount	Annual value

Maintenance services	1%	R$	3.836,00	R$ 46.030,00
Maintenance materials	3%	R$	11.508,00	R$138.090,00
General expenses	25%	R$	95.896,00	R$ 1.150.750,00
Insurance	2%	R$	7.672,00	R$ 92.060,00
Other fixed costs	1%	R$	3.836,00	R$ 46.030,00
*Operational and administrative labor		R$	54.850,00	R$658.200,00
TOTAL		R$	177.597,00	R$ 2.131.160,00

*Labor costs and expenses include social charges and other benefits.

General expenses include the cost of analyzing the quality of biodiesel, with a unit value of R$250.00. During the year, 54 analyses will be carried out, representing R$13,500.00.

4.2.3 Variable costs

Variable costs, unlike fixed costs, are those that vary in proportion to the level of industrial production and also to the sales structure adopted by the company. Obviously, what will determine whether a cost is fixed or variable is the nature of the company's activities and its production processes. The basic model adopted in this work for variable costs considers expenditure on raw materials, packaging, sales commissions and direct sales taxes.

The variable costs of the biodiesel production operation are based on the raw material used (babassu oil), electricity, alcohol and catalyst. The information listed in Table 25 was established on the basis of the manufacturer's budget, with the exception of babassu oil.

The cost of R$0.30 per liter of babassu oil was estimated on the basis of collecting and transporting the babassu coconuts to the cooperative, since in this project it was considered that the coconut processing would be done within the cooperative's facilities, as well as the oil extraction. In the project considered in this study, the coconut breakers, in addition to collecting the babassu coconuts, can extract the kernels and take them to the cooperative, receiving R$1.0 per kg of kernel extracted.

Table 25: Variable operating costs, inputs and unit price (12-hour period/25 days of the month).

Inputs	Unit price
Electricity	R$ 0.25/kWh

Babassu oil	R$ 0.30/liter
Catalyst	R$ 0.14/liter
Alcohol (methanol)	R$ 1.80/liter

Next, the variable costs were simulated according to the use of the plant. For this plant, each batch of 1000 L of biodiesel requires ~ 28.00 kg of catalyst (sodium methylate); ~ 1.1 L of babassu oil; ~ 210 L of anhydrous methanol; 230.00 liters of water, which results in the production of ~ 100.00 kg of crude glycerin. Based on the information obtained in the plant's budget proposal, an attempt was made to analyze the economic viability of the 300,000 L/month plant, as described in (Table 26). With regard to electricity, there was a change in the costs based on the commercial proposal due to its use in the babassu coconut processing machines. In the biodiesel plant proposal, every 2 hours the cost of electricity was estimated at R$25.00. Because of the other equipment, a value of R$75.00 was considered in the economic viability calculation. In the manufacturer's commercial proposal, production was estimated at 1,000 L every 2 hours, and based on this data, monthly and annual production was estimated.

Table 26: Variable costs of daily, monthly and annual biodiesel production (12-hour period/25 days of the month)* .

Description	Production: 1,000 liters of biodiesel (2 hours/day)*	Production: 300,000 liters of biodiesel per month (12 hours/day/25 days)	Production: 3. 600,000 liters of biodiesel per year
Babassu oil	R$ 316,00	R$ 94.800,00	R$ 1.137.600,00
Alcohol	R$ 378,00	R$ 113.400,00	R$ 1.360.800,00
Electricity for the plant and other equipment.	R$ 75,00	R$ 22.500,00	R$ 270.000,00
Catalyst	R$ 4,00	R$ 1.200,00	R$ 14.400,00
TOTAL	R$ 773,00	R$ 231.900,00	R$ 2.782.800,00

*Values calculated from the data provided in the commercial proposal of the manufacturer A.

4.2.4 Estimated revenues

The estimated revenues are shown in Table 27, assuming that all production is sold or disposed of according to the cooperative's interests.

The estimated biodiesel yield per liter of dry oil processed was 100% and the selling price was R$ 2.33/L, based on the average price of the 16th auction (ANP, 2009).

It was also considered that, along with the production of 1,000 liters of biodiesel, 100.00 kg of crude glycerin is obtained, the selling price of which is around R$ 0.50/kg and that this sale should be included in the billing projection.

The endocarp of the babassu shell, equivalent to 56%, used as the raw material for activated charcoal, will be sold at R$2.50 per 35 kg. Considering that it takes approximately 22 kg of babassu coconut to extract one liter of oil, and that 1.1 L of babassu oil is needed to transform one liter of oil into biodiesel, there will therefore be ~ 48,787,000 kg of babassu shell endocarp to produce 3,600,000 L of biodiesel per year. Therefore, the annual revenue from the endocarp corresponds to ~ R$ 3,484,800.00.

The transformation of the endocarp into activated charcoal was not considered in this research. To sell it, an additional investment will be needed, but it is a product that is well accepted in the market, being used by several companies, and the average purchase price is R$ 3.0 kg.

The cake, which corresponds to 2.4% of the babassu coconut, is sold for R$0.50/kg. The cooperative will have 2,090,880 kg of cake available each year, which will generate an annual income of R$1,045,440.00.

Table 27: Babassu products and by-products - annual value.

Product	Estimated production	Considered sale value, in R$	Estimated total sale value, in R$
Biodiesel	3,600,000 liters	2,33	8.388.000,00
Endocarp	48,787,000 kg	2.50 each 35kg	3.484.800,00
Pie	2,090,880 kg	0.50 Kg	1.045.440,00
Glycerin	360,000 kg	0.50 kg	180.000,00

In terms of tax rates, gross revenue from the sale of biodiesel by the producer or importer as of April 1, 2005, is subject to COFINS and PIS-Pasep at the following rates: PIS-Pasep: 6.15%; and COFINS: 28.32% (BIODIESELBR, 2005).

The rates for PIS/Pasep and COFINS (Social Contribution for the Financing of Social Security) on biodiesel production were set at 34.47%, but are reduced by 100% for family producers in the North and Northeast who produce biodiesel from castor beans or palm oil; 68% for family producers in any part of the country; and 32% for other producers. Contributions are levied on gross revenue and are 6.15% PIS/Pasep and 28.32% COFINS (BIODIESELBR, 2005).

In this work, the PIS/Pasep and COFINS rate of 11.03% was considered due to the 68% reduction from the stipulated 34.47%.

As for the ICMS, according to ICMS Agreement 113/06, the basis for calculating the Tax on Operations Relating to the Circulation of Goods and on the Provision of Interstate and Intermunicipal Transport and Communication Services - ICMS - is reduced, so that the tax burden is equivalent to 12% (twelve percent) of the value of the operations, on the outputs of biodiesel (B-100), resulting from the industrialization of grains, beef tallow, seeds, palm (BIODIESELBR, 2006a).

To calculate the taxes on the by-products of babassu coconut, the rates mentioned in Table 28 were used.

In the case of Income Tax on Net Profit - If net profit exceeds R$20,000.00 per month, there will be an additional 10% on the excess.

It can be seen that the tax burden for biodiesel is onerous, compromising profitability, which is why it is important to find a way to lower tax rates with the government to encourage production.

As biodiesel is sold at auction, the price could change, jeopardizing the cooperative's profits. However, revenues from babassu husks, cake and glycerin will have an impact on economic viability if biodiesel prices fall.

Table 29 shows that biodiesel accounts for 64.04% of gross operating revenue, babassu husk endocarp for 26.60%, cake for 7.98% and glycerin for 1.37%. Biodiesel is therefore the most profitable product, but the by-products are significant for adding value and meeting contingency needs. It can be seen that the babassu industrialization chain could grow even more as a result of research into this activity.

Government policies could help to reduce taxes, as biodiesel contributes to the Brazilian economy and the environment.

It is important to note that it was assumed that all the production obtained for biodiesel, babassu husk, cake and glycerin is effectively sold, obtaining maximum revenue.

Table 28: Tax rates for other products.

Type of Tax	Type of Activities	Rate	Calculation Basis
IR - Income Tax on Net Profit	Trade, Industry and Services	15%	Net Profit
CSLL - Social Contribution on Net Profit	Trade, Industry and Services	9%	Net Profit
PIS - Social Integration Program	Trade, Industry and Services	165%	Sale value
COFINS - Social Financial Contribution	Trade, Industry and Services	760%	Sale value
ICMS - Goods and Services Tax	Industry and Commerce	Variable from 0% to 25%	Sale value

Source: SEBRAE SP, 2010

Table 29: Projection of annual sales and revenues.

GROSS OPERATING REVENUE		13.098.240,00
PRODUCT SALES		
Biodiesel	R$	8.388.000,00
Babassu husk	R$	3.484.800,00
Pie	R$	1.045.440,00
Glycerin	R$	180.000,00
(-) DEDUCTIONS FROM GROSS REVENUE		3.215.297,00
ICMS - Biodiesel (12%)	R$	1.006.560,00
ICMS - Babassu bark, cake and glycerin (18%)	R$	847.843,00
PIS - Pasep and COFINS - Biodiesel (11.03%)	R$	925.197,00
PIS - Pasep and COFINS - Babassu bark, cake and glycerin (9.25%)	R$	435.697,00
(=) NET OPERATING REVENUE		9.882.943,00
(-) Variable costs	R$	2.782.800,00
(-) Fixed costs	R$	2.131.160,00
(-) Depreciation	R$	460.300,00
(=) PROFIT BEFORE SOCIAL CONTRIBUTION		4.508.683,00
(-) CSLL 9%	R$	405.782,00
(=) PROFIT BEFORE INCOME TAX		4.057.901,00
(-) INCOME TAX 15%	R$	676.303,00
(-) IR 10%	R$	426.868,00
(=) NET PROFIT	R$	3.404.730,00

4.2.5 Payback

This method aims to calculate the number of periods or how long the investor will need to recover the investment made. An investment means an immediate outflow of money. On the other hand, cash flows are expected to recover this outflow. Payback calculates how long this will take. This method is widely used by entrepreneurs to determine the attractiveness of an investment. Considering that the main objective of a project is profit and not the recovery time of the capital invested, this
method ignores any occurrence beyond the final period in which the capital was recovered.

Table 30 shows the payback calculation for the condition proposed in this study. It can be seen that the project is viable from an economic point of view if the plant operates with a 12-hour shift, 25 days a month for 1 year. The payback period is approximately 9 months.

Table 30: Payback for the condition proposed in this study.

DISCRIMINATION	MONTHS
Payback (Investment Value v Periodic Flow Value) Expected) (4.630.000,00 ÷ 3.404.730,00)	9

4.2.6 Financing using the National Program for Strengthening Family Farming - PRONAF Investment - ECO

Based on the BNDES Circular of October 13, 2009, the Credit Line for Investment in Renewable Energy and Environmental Sustainability - PRONAF ECO - is available.

Beneficiaries are Individuals who qualify as Family Farmers under PRONAF, and who submit a proposal or technical project for investments with the aim of implementing, using and/or recovering the items below:

a) Renewable energy technologies, such as the use of solar energy, biomass, wind power, biofuel plants and the replacement of fossil fuel technology with renewable technology in agricultural equipment and machinery.

b) Environmental technologies such as water, waste and effluent treatment plants, composting and recycling.

c) Water storage, such as the use of cisterns, dams, underground dams, water tanks and other storage and distribution structures, installation, connection and use of water.

d) Small hydro-energy projects.

e) Forestry, where forestry is understood to be the act of planting or maintaining forest stands that generate different products, both timber and non-timber.

f) Adopting conservation practices and correcting the soil's acidity and fertility, with a view to recovering it and improving its productive capacity.

The Effective Interest Rates for investment credits in the Conventional Line of PRONAF - ECO are from (BNDES, 2009):

- 1% p.a. (one percent per year), for one or more operations which, when added to the outstanding balance of "in being" financing, do not exceed R$ 7,000.00 (seven thousand reais).

- 2% p.a. (two percent per year) for one or more operations which, when added to the outstanding balance of "in being" financing, exceed R$ 7,000.00 (seven thousand reais) and do not exceed R$ 18,000.00 (eighteen thousand reais).

- 4% p.a. (four percent per year) for one or more operations which, when added to the outstanding balance of "in being" financing, exceed R$ 18,000.00 (eighteen thousand reais) and do not exceed R$ 28,000.00 (twenty-eight thousand reais).

- 5% p.a. (five percent per year) for one or more operations which, when added to the outstanding balance of "in being" financing, exceed R$ 28,000.00 (twenty-eight thousand reais) and do not exceed R$ 36,000.00 (thirty-six thousand reais).

The term for biofuel plant projects is up to 12 years, for collective operations the individual amount obtained by the proportionality of participation criterion is limited to R$ 18,000.00 (eighteen thousand reais), regardless of the limits set for other financing under PRONAF, in this criterion the rate is 2% p.a., and the amount per operation is limited to R$ 10,000,000.00.

Analyzing Table 31 with the objectives of the study, it can be seen that the amount of R$10,000,000.00 is vitally important for the acquisition of the biodiesel plant, working capital and the processing of the extraction of the kernels and the manufacture of the oil and the purchase of raw materials.

The rate of 2% p.a. (two percent per year) is attractive, and the 12-year term and 5-year grace period are also differentiators in the financial market.

This financing will have to be carried out collectively, for which approximately 556 cooperative members are needed, as shown in the line, the participation of each cooperative member is up to R$ 18,000.00.

In this project, the estimated annual net profit was R$3,404,730.00, divided among the 556 cooperative members, with each member receiving R$6,026.00 per year, and R$502.00 per month. Over the next few years, these figures could fall as a result of the loan repayments.

Simulating a loan amount of R$10,000,000.00, at a rate of 2%, with a grace period of 5 years and a term of 12 years, and considering the release date of 15/05/10, it can be seen that the first installment will be due on 15/05/2015 and the last on 15/05/2021 in the amount of R$1,584,540.00. The sum of all the financing installments is R$11,091,780.00. Therefore, a percentage of 10.92% will be paid on the loan amount.

Table 31: PRONAF line financing - Investment - ECO

Financing amount	10.000.000,00
Expiration date	5 years
Total Deadline	12 years old
Interest	2%
Rate	0,00
IOF (up to the grace period)	38.000,00
Release Dt	15/5/2010
Annual installments	1.584.540,00

4.2.7 Preferred uses for glycerol produced on a small scale

Of the possibilities associated with the use of glycerin in small biodiesel production units, three factors are of absolute and fundamental importance: the quality of the glycerin produced, the variation in its quality parameters over time and the logistics of its distribution to the local chemical industry. Firstly, if the quality of the glycerine is sufficiently acceptable and its specification is sufficiently constant over the time the plant is in operation, various industrial sectors will be able to benefit from it, such as initiatives to produce glycerinated soap, animal feed, composites for different applications and materials based on their composition with starch materials, as well as some types of direct use as a vehicle for dispersing agrochemicals, films for protecting surfaces from the elements, wetting agents for materials and surfaces, among others. Naturally, biodiesel producers can explore some of these initiatives themselves, using technologies that are already available on the market and/or in research institutes. However, as stated above, the quality of the glycerin produced and its maintenance throughout the plant's operating life are absolutely fundamental to increasing the pay-back on this important co-product of the biodiesel industry.

There is a lot of talk these days about burning crude glycerin to generate energy at the plant, thus increasing the unit's energy sustainability along with the use of other types of locally available residual biomass. However, it is known that the direct combustion of glycerin, under non-optimized conditions, can be extremely dangerous due to the formation of toxic aldehydes such as acrolein. Therefore, any initiative of this nature will necessarily have to be accompanied by professional guidance compatible with the activity it is intended to develop.

The proposal included in this research, regarding the supply of production units with a capacity in the range of 10,000 to 15,000 L/day, will lead to the accumulation of 1,000 to 1,500 liters per day of glycerin, which corresponds, at least theoretically, to 10% of the amount of oil processed. This is therefore a significant amount of glycerine, but not enough to justify large investments in its purification.

The economic viability of neutralizing, washing and concentrating crude glycerin on a small production scale has been hotly contested in scientific and business circles. For this reason, applications that can make use of crude glycerin appear to be the most viable. Naturally, the best option for disposing of glycerin would depend on a survey of local potential. For example, the existence of an industrial park in the vicinity, specialized in large-scale glycerol production, could easily absorb this surplus. It is really difficult to assess the profitability that such negotiations could offer for the investment, and this would certainly depend on the composition of the glycerin water generated in the process. Prospects could also be offered by other industrial activities, such as cement kilns, potteries, organic material recycling industries, among others.

In the specific case of the plant in question, assuming the challenge of disposing of crude glycerin locally and considering the demands of the Medio Mearim and Cocais regions, the following alternatives can be suggested:

(a) production and energy use of biogas from glycerin (anaerobic route);

(b) the use of glycerin in agricultural activities such as tobacco cultivation; and

(c) incorporation of crude glycerin into animal feed.

However, a number of other alternatives could emerge through local research, with the aim of properly disposing of an aqueous effluent whose accumulation in the environment could have disastrous environmental consequences.

CONCLUSIONS

Babassu was the raw material selected in this study because it is abundant in the Médio Mearim and Cocais regions, which neighbor each other, and contributes to socio-economic activities. The population living in the Cocais and Médio Mearim regions is young and could be used as labor in the production of biodiesel and by-products.

The Middle Mearim region was chosen in this study to install the plant, due to the existence of ASSEMA in the region, which in turn has cooperatives, an essential factor for collective financing in the PRONAF line with a limit of up to R$ 10,000,000.00.

The economic viability calculation showed that the proposal presented by manufacturer A is viable for a 12-hour shift operation, 25 days a month. Within this analysis, it was found that the estimated annual net profit was R$3,404,730.00 with a payback of approximately 9 months.

Because biodiesel is sold at auction, the price could change, jeopardizing the cooperative's profits. However, the income from babassu shells, cake and glycerin will contribute to economic viability if the price of biodiesel falls.

With the installation of the enterprise planned in this study, each cooperative member will receive R$6,026.00 per year, and R$502.00 per month, approximately five hundred and fifty-six families will benefit. Over the next few years, these amounts could decrease as a result of the loan repayments. Taking into account that the majority of the population has an income of between 0.5 and 2 minimum wages and based on the current minimum wage of R$510.00, the increase in income for cooperative members of 0.5 minimum wages will be approximately 100% and 49% for 2 minimum wages, assuming that the income from the cooperative is added to the income they currently receive.

The total investment needed to set up the enterprise is R$4,630,000.00. As the loan was for R$10,000,000.00, R$5,370,000.00 could be used to expand the infrastructure for the babassu production chain, such as the production and sale of activated charcoal, the production of soaps and cleaning products with glycerin. This could further improve the return on investment and repayment of the loan.

In short, the installation of a plant in the state of Maranhão, more specifically in the Médio Mearim region, using babassu oil and small farmers organized in cooperatives, is socio-economically viable.

SUGGESTIONS FOR FUTURE WORK

- Analysis of the expansion of the babassu by-product chain.
- Development of an information and management system so that cooperatives can manage their processes more quickly and accurately.
- Evaluate the possibility of adding other oilseeds to the region where the babassu trees are located, expanding the biodiesel production capacity.
- Elaboration of an educational development project through technical and vocational schools, improving activities in the regions of Maranhão with a focus on entrepreneurship.
- Analysis of the feasibility of exporting biodiesel from Maranhão.
- Development and improvement of machines and implements for harvesting and processing babassu coconut.
- Development of public policies to encourage the activities of coconut breakers and small farmers in the Médio Mearim and Cocais regions.

BIBLIOGRAPHICAL REFERENCES

ABIOVE - BRAZILIAN ASSOCIATION OF VEGETABLE OIL INDUSTRIES. Seminar on Biodiesel in Rio Grande do Sul. 2005, Canoas: Refap, May 2005.

ABIQUIM - BRAZILIAN CHEMICAL INDUSTRY ASSOCIATION. Yearbook of the Brazilian Chemical Industry - 2007. São Paulo: ABIQUIM, 2008.

AGUIAR, A. C. S. Diesel demand and prospects. First meeting on vegetable oil fuel technology. In: Energia, 11, v. 2, pag. 11, 1980. AMERICAN SOCIETY FOR TESTING METHODS - ASTM. D6751. USA, 2003.

ALBIERO, D; MACIEL, S.J.A; LOPES, C. A; MELLO, A. C; GRAMERO, A . C. Proposal for a mechanized babassu harvesting machine for family farming. Acta Amazônica. Volume 37. Manaus, 2007.

ANP. Preliminary specifications for biodiesel in Brazil - Resolution No. 07. Available at < www.anp.gov.br> Accessed on 19/09/2008.

ANP. 16th biodiesel auction. Available at < www.anp.gov.br> Accessed on 13/12/2009.

ASSEMA. Association in settlement areas in the state of Maranhão. Available at < www.assema.org.br> Accessed on 12/10/2009.

ASSUMPÇÃO, R. M. Master's thesis. "Market vision on the availability of raw materials for biodiesel production. A case study in the state of Paraná", 119, pg, 2006.

BAILEY, A. E.; HUI, Y. H. Bailey's industrial oil and fat products. 5 ed. New York: John Wiley, 2005. v. 5, 275-308 p.

BANCO DO BRASIL S.A. - Banco do Brasil Program to Support the Production and Use of Biodiesel. Available at:< http://www.agronegocios- e.com.br/agronegocios> Accessed on August 5, 2006.

BEGNIS, Heron et al. Integrated Production in Cooperative Agro-Industrial Organizations: a case study based on the dynamics of complex systems. Estudos do CEPE, Santa Cruz do Sul, n.20, p.59-80, jul/dez. 2004.

BEZERRA, O. B. Localization of collection points to support the flow of extractive products: a case study applied to babassu. Santa Catarina: UFSC, 1995. 100 p

BIALOSKORSKI NETO, Sigismundo. Cooperative Agribusiness. In: Zylberstajn, Decio; NEVES, Marcos Fava. Economia e Gestão dos Negócios Agroalimentares. São Paulo: Pioneira Thomson Learning, 2005.

BIODIESEL (2006). Social Fuel Seal . Available at: < http://www.biodiesel.gov.br/selo.htm> Accessed on 10/12/2008.

BIODIESEL (2009). National Program for the Production and Use of Biodiesel - PNPB. Available at: < http://www.biodiesel.gov.br/programa.html> Accessed on 18/04/2009.

BIODIESELBR (2005). COFINS AND PIS - Pasep - Biodiesel - Applicable rates. Available at : <http://www.biodieselbr.com>. Accessed on: 12/08/2009.

BIODIESELBR (2006). Transesterification: Details on the stages of biodiesel production. Available at : <http://www.biodieselbr.com>. Accessed on: August 22, 2009.

BIODIESELBR (2006a). Biodiesel calculation base reduced to 12%. Available at : <http://www.biodieselbr.com>. Accessed on: October 07. 2009.

BIODIESELBR (2009). With more biodiesel, there is more glycerin and the risks are growing. Available at : <http://www.biodieselbr.com>. Accessed on: December 12, 2009.

BNDES. National Program to Strengthen Family Farming - PRONAF. Available at < http://www.bndes.gov.br>. Accessed on December 10, 2009.

BNDES (a). Cooperative Development Program for Adding Value to Agricultural Production - PRODECOOP. Available at < http://www.bndes.gov.br>. Accessed on December 10, 2009.

BRAZIL. Ordinary Law No. 5764 of December 16, 1971. Defines the national cooperative policy, establishes the legal regime for cooperative societies and makes other provisions. Brasília: Federal Official Gazette, 1998.

CANÇADO, Airton Cardoso; GONTIJO, Mário César Hamdan. Cooperative principles: origin, evolution and influence on Brazilian legislation. III Meeting of Latin American Cooperative Researchers. UNISINOS, São Leopoldo: April 28, 29 and 30, 2004.

CARDIAS, C .T . H. Biodiesel program in Maranhão. Brasília: Ministry of Science and Technology, 2005.

CASTELANI, A. C. Estudo da Viabilidade de Produção do biodiesel, obtido através do óleo de fritura usado, na cidade de Santa Maria - RS. Master's dissertation. Santa Maria: UFSM, 2008.

CEPLAC. Oil palm potential for renewable energy production. Available at: http://www.ceplac.gov.br/radar/Artigos/artigo9.htm. Accessed on 29/09/2009.

CHIARANDA, M.; ANDRADE JÚNIOR, A. M.; OLIVEIRA, G. T. Biodiesel production in Brazil and aspects of the PNPB. Piracicaba: ESALQ/USP, 2005 (GEEDES Research Report/Department of Economics, Administration and Sociology).

CLEMENT, C. R.; Lleras Peres, E.; van Leeuwen, j. 2005. The potential of tropical palms in Brazil: successes and failures of recent decades. Agrociencias, 9, (1-2): 67-71.

CONCEIÇÃO, M. M.; Jr, V. J.; SINFRÔNIO, F. S. M.; SANTOS, J. C. O.; SILVA, M. C. D.; FONSECA, V. M.; SOUZA, A. G.; Journal of Thermal Analysis and Calorimetry, 461 (79), 2005.

COSTENARO, Sonego Hellen. Acid hydrolysis removed from crude glycerin salts from biodiesel production. Master's thesis. São Paulo, USP, 2009.

DEMIRBAS, M.F., BALAT, M. Recent advances on the production and utilization trends of bio-fuels: A global perspective. Energy Conversion and Management, Volume 47, Number 15-16, p. 2371-238, 2006.

DEUTSCHES INSTITUT FÜR NORMUNG E.V. - DIN, EN14214. Germany, 2003.

EPE. Ten-Year Energy Expansion Plan 2008 - 2017. Ministry of Mines and Energy. Energy Research Company. Rio de Janeiro: EPE, 2009.

EET CORPORATION. Glycerin Purification. Available at: http://www.eetcorp.com/heepm/glycerine.htm. Accessed on March 20, 2009.

FANGRUI, M.; HANNA, M. A.; Bioresource Technology 70: 1, 1999.

FREEDMAN, B.; PRYDE, E. H., MOUNTS, T. L.; Journal American Oil Chemists Society, 61 (10): 1638, 1984.

GOLDEMBERG, J., TEIXEIRA COELHO, S., NASTARI, P.M., LUCON, O. Ethanol learning curve the Brazilian experience. Biomass and Bioenergy, Volume 26, Number 3, p. 301-304, 2004.

FILHO, Afrisio Vieira Lima. Biodiesel and social inclusion. Oct. 2003. Available at: http://www.sfiec.org.br/artigos/tecnologia/BIODIESEL_2003.pdf> Accessed on: Dec. 2009.

FRAZÃO, J. M. F. 2001. Economic alternatives for family farmers settled in areas of babassu ecosystems. Technical report. Maranhão State Government, São Luis. 120pp.

GIL, Antonio Carlos. How to prepare research projects. São Paulo: Atlas, 1991.

HATEKEAMA, T.; QUINN, F. X.; Thermal analysis, John Willey & Sons, Japan, 1994.

HERRMANN, I.; NASSAR, A. M.; MARINO, M. K. M.; NUNES, R. Coordination of the Babassu SAG: Is rational exploitation possible? < http://www.pensa.org.br/Biblioteca.aspx?tipo=10 > . Accessed on 20/12/2009.

HOLANDA, A. Rapporteur. Biodiesel and Social Inclusion. Chamber of Deputies. High Studies and Technological Assessment Council. Videoconference. Brasília: 2004.

IMESC (a). Profile of the Médio Mearim Region / Instituto Maranhense de Estudos Sócio - econômicos e Cartográficos. São Luís : IMESC, 2009.

IMESC (b). Profile of the Cocais Region / Instituto Maranhense de Estudos Sócio - econômicos e Cartográficos. São Luís : IMESC, 2009.

PLASTICS NEWSPAPER. Nova Petroquímica is the first in the world to develop polypropylene plastic with biodiesel derivatives. March 2008. Available at: http://www.jorplast.com.br/jpmar08/pag02.html. Accessed on May 15, 2008.

KAPLAN, S. Energy Economics Quantitative Methods for Energy and Environmental Decision. Mac Graw Hill, 1983.

KNOTHE, G.; KRAHL, J.; VAN GERPEN, J. The Biodiesel Handbook. Illinois: AOCS Press, 2005.

KNOTHE, G.; STEIDLEY, K. R.; Fuel 84: 1059, 2005.

KNOTHE, G.; GERPEN, J. V.; KRAHL, J.; RAMOS, L.P. Manual de Biodiesel, Ed. Edgarb Blucher, 2006.

LADETEL. LABORATORY FOR THE DEVELOPMENT OF CLEAN TECHNOLOGIES (LADETEL / USP-RP). Biodiesel: strategies for production and use in Brazil. In: BIODIESEL: STRATEGIES FOR PRODUCTION AND USE IN BRAZIL, 2005, São Paulo: Unicorp, April 26-27, 2005. Proceedings... v.1, p. 34-37.

LIMA, P. C. R. Biodiesel: A new fuel for Brazil. Brasília (DF): Consultoria Legislativa, Feb. 2005.

LOPES, O.C.; Schuchardt, U.F.F. 1983. New catalysts for the transesterification of vegetable oils. Master's dissertation, Faculty of Agricultural Engineering, State University of Campinas, Campinas, SP. 1983. 103pp.

LOPES, O.C.; MACIEL, A.J.S., Method of transesterification of vegetable oils and animal fats, catalyzed by modified strong base for the production of Biodiesel, 2005. Patent: Patent and Invention No. 050 231 2 -2 filing date June 2005.

MA, F.; HANNA, M. A. Biodiesel production: a review. Bioresorce Technology, 1999, v. 70, p. 1-15. MAPA. National Agroenergy Plan. Brasília, Oct. 2005. Available at: < http://www.agricultura.gov.br > Accessed on 13/10/2008.

MARANHÃO. Bioenergy - Investment opportunities. Maranhão. Maranhão State Government, 2008.

MAY, P.; Veiga Neto, F.C.; Cheves Pozo, O.V. 2000. Economic valuation of biodiversity. Ministry of the Environment, Brasília. 90pp.

MAX S.P. and D.T. KLAUS. Plant Design and Economics for Chemical Engineers, 4ed., McGraw-Hill, Inc., USA, 1991.

MDIC - MINISTRY OF DEVELOPMENT INDUSTRY AND FOREIGN TRADE. Brazil exporter. Available at:

http://www.desenvolvimento.gov.br. Accessed on March 25, 2008.

MELLO, T.O.F.; PAULILO, F. L.; VIAN, F. E C. Biodiesel in Brazil: panorama, prospects and challenges. Informações econômicas, SP, v 37, n1, January 2007.

MITTELBACH, M.; TRITTHART, P. Diesel fuel derived from vegetable oils, II: emission tests using rape oil methyl Ester. Energy in Agriculture, 1985, v.4, p.207-215.

MIQCB. Interstate Movement of Babassu Coconut Breakers. Available at: < http://www.miqcb.org.br/galeria.html > Accessed on 20/10/2009.

MME. MINISTRY OF MINES AND ENERGY.Ten-Year Energy Plan 2008 to 2017 . Energy Research Company. Rio de Janeiro: EPE, 2009

MME. MINISTRY OF MINES AND ENERGY. B5 becomes mandatory as of January 1, 2010. <http:// www.mme.gov.br > Accessed on 02/01/2010.

MRE - Ministry of Foreign Affairs. The State of Maranhão. Available at http://www.mre.gov.br/dc/textos/revista2-mat10.pdf. Accessed on October 20, 2009.

NASCIMENTO, U. S. Babassu charcoal as a thermal source for absorption refrigeration systems in the state of Maranhão. 2004. 99 p. Dissertation (Master's Degree in Mechanical Engineering) - Postgraduate Course in Mechanical Engineering, State University of Campinas, 2004

NATIONAL BIODIESEL BOARD: Proceedings of the International Congress on Liquid Biofuels; Paraná Institute of Technology; Secretary of State for Science, Technology and Higher Education; Curitiba, Brazil: July 19-22, 1998, p.2.

NYE, M. J.; WILLIAMSON T.W.; DESHPANES, E.; SCHRADER J. H.; SNIVELY, W. H.; YURKEWICH T. P.; FRENCH C. L; Journal of the American Oil Chemists Society 60 (8): 1598, 1983.

OCB - Organization of Brazilian Cooperatives. Brazilian cooperatives. Available at: http://www.ocb.org.br. Accessed on: May 23, 2009.

OLIVEIRA, Djalma de Pinho Rebouças de. Manual de gestão das cooperativas - uma abordagem prática. São Paulo: Atlas, 2001.

PAGLIARDI, O; J.M.MESA, J.D. ROCHA, E. OLIVARES , L.A BARBOZA. Biomass Fast Pyrolysis Plant: Aspects of Economic Viability. IV Brazilian Congress of Energy Planning, Itajubá (MG), March 2004.

PARENTE, E.J. de S.; SANTOS JUNIOR, J.N.; PEREIRA, J.A.B.; PARENTE JUNIOR, E.J.de S. Biodiesel: a technological adventure in a funny country. Fortaleza: Tecbio, 68p. 2003.

PENSA. Reorganization of the Babassu Agribusiness in the State of Maranhão. USP, 2000. In: SANTOS, N. A. Thermo-oxidative and flow properties of babassu biodiesel (Orbignya phalerata). 2008. Master's dissertation. Federal University of Paraíba, João Pessoa, PB.

PENTEADO, M.C.P.S. Identification of Bottlenecks and Establishment of an Action Plan for the Success of the Brazilian Biodiesel Program. São Paulo, 2005, 159 f. Dissertation (Master's Degree) - Automotive Engineering, Polytechnic School of USP, São Paulo, 2005.

PETROBIO. Biodiesel: Economic viability. 24 p., 2005.

PINHO, Diva Benevides. O cooperativismo no Brasil - da vertente pioneira à vertente solidária. São Paulo: Saraiva, 2004

PINTO, E.; MENDONÇA, M. M. The fuel myth . Brasil de Fato. Available at: <http://www.brasildefato.com.br>. Accessed on December 19, 2009.

PLANO NACIONAL DE AGROENERGIA 2006-2011/MINISTÉRIO DA AGRICULTURA, PECUÁRIA E ABASTECIMENTO, SECRETARIA DE PRODUÇÃO E AGROENERGIA. 2. ED.REV. - BRASÍLIA, DF: EMBRAPA INFORMAÇÃO TECNOLÓGICA, 2006.

NOVAES PLANTS. Babassu coconut. Available at

http://www.plantasnovaes.com/Galeria/images/babacu.jpg. < Accessed on 04/12/2009.

PORTO, M. J. F. Preliminary study of a breaking device and characterization of the physical parameters of babassu coconut. 2004. 75 f. Dissertation (Master's in Mechanical Engineering), Campinas State University, 2004.

PRATES, C.P.T.; PIEROBON, E.C.; COSTA, R.C.; Formation of the biodiesel market in Brazil. BNDES Setorial, Rio de Janeiro, n.25, p. 39-64, mar. 2007.

QUEIROZ, Carlos Alberto Ramos Soares de. Manual da Cooperativa de Serviços e Trabalho. São Paulo: Editora STS, 1998.

RAMOS, L. P. Converting vegetable oils into an alternative biofuel to conventional diesel. In: BRAZILIAN SOY CONGRESS, Londrina, 1999. Proceedings. Londrina: Embrapa-Soja, 1999, p. 233-236.

RAMOS, L. P.; DOMINGOS, A. K.; KUCEK, K. T.; WILHELM, H. M. Biodiesel: An economic and socio-environmental sustainability project for Brazil. Biotecnologia: Ciência e Desenvolvimento, 2003, v.31, p.28-37.

RAMOS, L.P.; WILHELM, H.M. Current status of biodiesel development in Brazil. Applied Biochemistry and Biotechnology, 2005, v. 121-124, p. 807-820.

RICHARDSON, Roberto Jarry et al. Pesquisa Social: Métodos e Técnicas. São Paulo: Atlas, 1985.

SALVADOR, A.F., MACHADO, A.S., SANTOS, E.P. Purificação da Glicerina Bruta Vegetal. Biodiesel: the country's new fuel. SENAI-CETIND, 2006. 20-23 p.

SEBRAE SP. Taxes levied on the company's operational activities. Available at: http://www.sebraesp.com.br/faq/financas/legislacao/ impostos_atividades_operacionais. Accessed on: 20/01/2010.

SEPLAN. Planning regions of the state of Maranhão / State Secretariat for Planning and Budget. Maranhão Institute of Socio-Economic and Cartographic Studies. State University of Maranhão. São Luiz, 2008.

SERIO, M. DI; LEDDA, M.; COZZOLINO, M.; MINUTILLO, G.; TESSER, R.; SANTACESARIA, E. Transesterification of Soybean Oil to Biodiesel by Using Heterogeneous Basic Catalysts. Ind Eng. Chem. Res. v. 45, 2006, p. 3009-3014.

SOUZA, A. S. Biodiesel and vegetable oils as an alternative for electricity generation: the positive example of Rondônia. In: GREENPEACE. (Positive dossier for BRAZIL. Available at: <http://www.greenpace.org.br>. Accessed on Dec. 20, 2009.

TRIVIÑOS, Augusto N. S. Introdução à Pesquisa em Ciências Sociais: a pesquisa qualitativa em

Educação. São Paulo: Atlas, 1987.

ULLMANN, F.et al. Ullmann's encyclopedia of industrial chemistry , completely rev. 5ºed .1988. 477-489 p.

UOL. Child labor in the babaçuais. Available at http://n.i.uol.com.br/ultnot/0909/17crianca2.jpg > . Accessed on 20/12/2009.

VAN GERPEN, J.; KNOTHE, G. Basics of the transesterification reaction. In: KNOTHE, G.; KRAHL, J.; VAN GERPEN, J. (eds.) The Biodiesel Handbook. Illinois: AOCS

VARGAS, R. M.; SCHUCHARDT, U.; SERCHELI, R.; Journal Brazilian Chemists Society, 9 (1): 199, 1998.

VERGARA, Sylvia Constant. Research projects and reports in administration. 3. ed. São Paulo: Atlas, 2000.

WASSELL JR., C.S. & DITTMER, T.P. Are subsidies for biodiesel economically efficient? Energy Policy, Volume 34, Issue 18, December 2006, Pages 3993-4001.

WILHELM, H, M; PORTELA, F. K; RAMOS, P.L; TULIO, L.; KERECZ, A. Survey of the technical, economic and social feasibility of setting up a biodiesel production unit in Maranhão. Curitiba: Lactec, 2008.

WUNDER, S. Value determinants of plant extractivism in Brazil. Center for Development Research, Copenhagen, 1998. 210pp

ZYLBERSZTAJN, D.; Marques, C. A. S.; Nassar, A. M.; Pinheiro, C. M.; Martinelli, D. P.; Adeodato S. Neto, J.; Marino, M. K.; Nunes, R. Reorganization of the babassu agribusiness in the state of Maranhão. Technical report. Pensa-USP Group, São Paulo, 2000. 120pp.

MIX
Papier aus verantwortungsvollen Quellen
Paper from responsible sources
FSC® C105338

Printed by Books on Demand GmbH, Norderstedt / Germany